CHARLES WILLIAM ANDREWS

A DESCRIPTIVE CATALOGUE

OF THE

MARINE REPTILES

OF

THE OXFORD CLAY

Part I

Elibron Classics
www.elibron.com

Elibron Classics series.

© 2006 Adamant Media Corporation.

ISBN 0-543-94108-6 (paperback)
ISBN 0-543-94107-8 (hardcover)

This Elibron Classics Replica Edition is an unabridged facsimile
of the edition published in 1910 by the British Museum, London.

Elibron and Elibron Classics are trademarks of
Adamant Media Corporation. All rights reserved.

This book is an accurate reproduction of the original. Any marks, names, colophons, imprints, logos or other symbols or identifiers that appear on or in this book, except for those of Adamant Media Corporation and BookSurge, LLC, are used only for historical reference and accuracy and are not meant to designate origin or imply any sponsorship by or license from any third party.

FRONTISPIECE

CATAL. MARINE REPT. OXFORD CLAY.

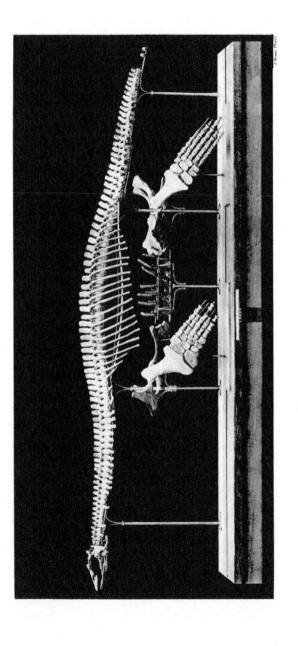

CRYPTOCLEIDUS OXONIENSIS.

A DESCRIPTIVE CATALOGUE

OF THE

MARINE REPTILES

OF

THE OXFORD CLAY.

BASED ON THE LEEDS COLLECTION IN
THE BRITISH MUSEUM (NATURAL HISTORY), LONDON.

PART I.

BY

CHARLES WILLIAM ANDREWS, D.Sc., F.R.S.

LONDON:
PRINTED BY ORDER OF THE TRUSTEES OF THE BRITISH MUSEUM.

SOLD BY
LONGMANS & Co., 39 PATERNOSTER ROW, E.C.;
B. QUARITCH, 11 GRAFTON STREET, NEW BOND STREET, W.; DULAU & Co., Ltd., 37 SOHO SQUARE, W.;
AND AT THE
BRITISH MUSEUM (NATURAL HISTORY), CROMWELL ROAD, S.W.

1910.

(All rights reserved.)

PRINTED BY TAYLOR AND FRANCIS,
RED LION COURT, FLEET STREET.

PREFACE.

During the past twenty years the British Museum has gradually acquired the fine collection of Reptilian skeletons obtained by the Messrs. Leeds, of Eyebury, from the Oxford Clay in the neighbourhood of Peterborough. Most of the specimens represent marine Reptiles of the Orders Ichthyopterygia, Sauropterygia, and Crocodilia; and the associated sets of bones have been extricated from the rock with so much skill and care that they afford an unique opportunity for acquiring a good general knowledge of the Reptilian fauna existing in the Upper Jurassic sea. Dr. Charles W. Andrews has therefore been entrusted with the preparation of a Descriptive Catalogue of the collection, and it is hoped that his exhaustive work will form a useful basis for future researches in the same field. The separate bones of many of these reptiles have now been studied and described as thoroughly and satisfactorily as if they were from freshly-macerated skeletons; and it is only to be regretted that a considerable proportion of the specimens are too much distorted by crushing in the soft moist clay to allow of any exact measurements. The variations observed in the different individuals of some species are especially noteworthy; and the growth-stages traceable in certain parts, such as the Elasmosaurian shoulder-girdle, are also of great interest. In accomplishing his task Dr. Andrews has been much assisted by Mr. Alfred N. Leeds, who made the greater part of the collection, and has given the Museum the benefit of his long experience.

Part I. contains the account of the Ichthyosaurs and Plesiosaurs. Part II. will be devoted to the Pliosaurs and Crocodiles.

A. SMITH WOODWARD.

Department of Geology,
British Museum (Natural History),
13th July, 1910.

INTRODUCTION.

NEARLY all the remains of the marine Reptilia of the Oxford Clay enumerated and described in this Catalogue, were collected from the numerous clay-pits near Peterborough worked for the making of bricks, an industry that is extensively carried on in that neighbourhood. A few of the earlier specimens were discovered by Mr. Charles E. Leeds, M.A., but the greater part of the collection was made by his brother, Mr. Alfred N. Leeds, F.G.S., of Eyebury, who soon became associated with him. It is more than forty years since the collection was begun by Mr. Charles E. Leeds, and some of his first discoveries were described and figured by Phillips in his 'Geology of Oxford and the Valley of the Thames,' published in 1871. He left for New Zealand in 1887, but his brother has continued the work to the present day with the most astonishing results. Both in the number of species represented and in the perfect preservation of their remains, the Leeds Collection far surpasses any other single collection of Mesozoic Vertebrates, especially one in which all the specimens are from one horizon and from a restricted area. Not only marine forms, but remains of terrestrial reptiles, including several species of Dinosaurs, have been obtained.

In nearly all cases the specimens have been collected with extreme care, usually by Mr. Leeds himself, the bones of the different parts of the skeleton being numbered and packed in separate parcels. Frequently, portions of the skeleton, such as the skull or limb-girdles, can only be extricated from the clay in fragments, but these have been reunited with the greatest skill and patience by Mr. Leeds. The consequence of this care is that, in the case of some of the more nearly complete and uncrushed skeletons, it has been possible to mount the bones in their natural relations as easily as if they had been obtained by the maceration of a fresh carcass. A notable instance of this is the fine skeleton (R. 2860) of *Cryptocleidus oxoniensis*, which is figured on the Frontispiece and forms the basis of the restoration given in text-figure 94 on page 188. In this case, as in many others, the bones, which all belong to a single individual, are uncrushed and undistorted. Often, however, the skeletons have been subjected to great pressure, and have thus been extensively fractured and deformed. Unfortunately the skulls are especially liable to injury, and therefore any specimens approaching completeness are very rare. Occasionally the whole skeleton or portions of it are embedded in an intensely hard pyritous clay, and when this is the case all attempts at clearing away the matrix are usually hopeless. Another cause of the imperfection of many of the skeletons seems to have been the dismemberment of the carcasses by carnivorous reptiles, probably some of the Crocodiles and Pliosaurs whose remains are also common in the Oxford Clay. Bones are often found scored across by deep grooves, obviously cut by sharply pointed teeth. Moreover, the curious manner in which whole sections of the skeleton, as, for example, a limb, are sometimes wanting in otherwise nearly complete specimens, or, on the other hand, the occurrence of isolated paddles and other parts of the skeleton, seems to show that the dismemberment occurred while the bones were still united by the soft tissues. A notable instance of this incompleteness is seen in the case of the portions of the skeleton of the giant Dinosaur, *Cetiosaurus leedsi*, described and figured by Dr. A. S. Woodward (Proc. Zool. Soc. 1905). In this specimen, the skull, the whole of the vertebral column in front of the sacral region, the left fore limb and the right hind limb, as well as the ischia and pubes, are wanting, while the left hind limb is almost complete, even to the phalanges, also the right fore limb except the manus; the vertebral column of the tail, again, is represented by two series of successive and complete vertebræ, an anterior series of about 27, and a posterior one of 10, while the intervening portion is entirely absent. Although careful search and extensive excavations have been made, none of the missing parts have been found, and such absence of whole sections of the body seems to be best explained by supposing the carcass to have been dismembered while the bones were still united.

The horizon at which these reptilian bones occur is that characterised by the presence of the "*Ornatus*" group of Ammonites, one of the species most commonly found in actual association with the bones being *Cosmoceras gulielmii*,

J. Sowerby (=*Ammonites jason*, Reinecke, sp., *fide* Oppel). Many other species of Cephalopods have been collected in the beds by Mr. Thurlow Leeds and others, the most important being *Cardioceras lamberti*, *C. serratum*, *Cosmoceras spinosum*, *C. duncani*, *C. ornatum*, *Peltoceras athleta*, *P. williamsoni*, var., *Aspidoceras perarmatum*, *Quenstedtoceras mariæ*, and *Belemnites oweni*.

The horizon at which these forms occur is described by English stratigraphers as the Lower Oxford Clay *. By continental geologists strata of the same age would be called Upper, or Middle and Upper Callovian, but, as Mr. H. B. Woodward † has remarked in the memoir referred to below, "This seems a quite unwarranted stretching of a formation to suit local stratigraphy and in defiance of its original significance." Accordingly the horizon in which the reptilian remains are found is here called the Lower Oxford Clay (Middle Oxfordian). The general succession of the beds of this age in Northamptonshire has been described by Prof. J. W. Judd ‡ under divisions *b–e* as follows:—

(*f*) Zone of *Ammonites cordatus*.

(*e*) Clays with Ammonites of the group of the *Ornati*.
   Dark blue clays with nodules of pyrites and numerous pyritic Ammonites, including *A. ornatus*, *A. duncani*, *A. bakeriæ*, and *A. athleta*, and also *Waldheimia impressa*.
   Dug in brickyards about Whittlesey, at Thorney, and Eye Green.

(*d*) Clays with *Belemnites hastatus*.
   Blue clays with many fossils found in Division *c*, but characterised by the abundance of *B. hastatus*.
   Dug at Werrington, Ramsey, and Eyebury.

(*c*) Clays with *Belemnites oweni*.
   Dark blue clays and shales with *B. oweni*, often of gigantic size. *Gryphæa dilatata* occurs, but is more plentiful in the beds above. Saurians and fishes occur, and masses of lignite, sometimes converted into jet, are found.
   Exposed in brickyards at Standground, Fletton, and Woodstone, near Peterborough, and at Connington, Luddington, and Great Gidding.

(*b*) Clays with *Nucula*.
   Laminated blue shales with compressed Ammonites and *Nucula nuda*.
   Dug at Haddon, Holme, south of Peterborough, and at Eyebury to the north-east.

(*a*) Zone of *Ammonites calloviensis*.

---

* See 'Memoirs of Geol. Survey of the United Kingdom—The Jurassic Rocks of Britain,' vol. v. "The Middle and Upper Oolitic Rocks of England," by H. B. Woodward (1895) p. 8.
† *Tom. cit.* p. 9.   ‡ 'Geology of Rutland' (1875) p. 232.

There seems to be a little uncertainty as to the conditions under which these beds were deposited. They are usually supposed to have been laid down in fairly deep water, but the presence of remains of land-reptiles, and perhaps the occurrence of large masses of lignite, may indicate that the coast was not far off. Probably they were mud-banks accumulating off the mouth of a large stream.

In the Collection at the British Museum (Nat. Hist.) there are remains of the following Vertebrates collected from the series of clays just enumerated:—

REPTILIA.

Cetiosaurus leedsi, Hulke, sp.
Omosaurus durobrivensis, Hulke.
A large Stegosaurian.
Camptosaurus leedsi, Lydekker.
Sarcolestes leedsi, Lydekker.
Rhamphorhynchus sp.
Ophthalmosaurus icenicus, Seeley.
Murænosaurus leedsi, Seeley.
    ,,   durobrivensis, Lydekker.
    ,,   platyclis, Seeley.
Picrocleidus beloclis, Seeley, sp.
    ,,   sp.
Tricleidus seeleyi, Andrews.
Cryptocleidus oxoniensis, Phillips, sp.

Pliosaurus ferox, Sauvage, sp.
Simolestes vorax, Andrews.
Peloneustes philarchus, Seeley.
Metriorhynchus superciliosus, Deslongchamps.
    ,,   brachyrhynchus, Deslongchamps.
    ,,   sp.
Suchodus durobrivensis, Lydekker.
Dacosaurus sp.
Steneosaurus edwardsi, Deslongchamps.
    ,,   leedsi, Andrews.
    ,,   nasutus, Andrews.
    ,,   durobrivensis, Andrews.
    ,,   obtusidens, Andrews.

It is possible that the number of species of Crocodiles may be increased when the material is examined later in detail. It is a very remarkable circumstance that no trace of any Chelonian has been found.

PISCES.

Hybodus obtusus, Agassiz.
Asteracanthus ornatissimus, Agassiz, var. flettonensis, A. S. Woodward.
Pachymylus leedsi, A. S. Woodward.
Brachymylus altidens, A. S. Woodward.
Ischyodus egertoni, Buckland, sp.
    ,,   beaumonti, Egerton.
Lepidotus leedsi, A. S. Woodward.
    ,,   latifrons, A. S. Woodward.
    ,,   macrocheirus, Egerton.

Heterostrophus sp.
Mesturus leedsi, A. S. Woodward.
Caturus sp. 1.
    ,,   sp. 2.
Osteorachis leedsi, A. S. Woodward.
Eurycormus egertoni, Egerton, sp.
Hypsocormus leedsi, A. S. Woodward.
    ,,   tenuirostris, A. S. Woodward.
Leedsia problematica, A. S. Woodward.
Pholidophorus sp.

INTRODUCTION.

Most of the fish-remains are important as exhibiting osteological characters which cannot be seen so satisfactorily in crushed specimens preserved in hard rock.

In this volume only the Ichthyosaurs and the Elasmosaurian Plesiosaurs are dealt with, the Pliosaurs and Crocodiles being reserved for the second volume.

The Ichthyosaurs are represented by one genus, *Ophthalmosaurus*, of which only a single species, *O. icenicus*, Seeley, is here recognised, though the variability of the skeleton is so great that some might be inclined to consider several species to be present. It has, however, been found, from the examination of a very large number of more or less nearly complete skeletons, that the different forms pass into one another, so that no line between this and that can be drawn. Many of the apparent differences are due to the different extent to which ossification has proceeded in individuals of various ages, and others arise from the differing conditions of preservation (presence or absence of compression, &c.).

Ophthalmosaurus, which is here regarded as congeneric with *Baptanodon* * from contemporary or approximately contemporary deposits of the United States, seems in many respects to represent the most highly specialised type of Ichthyosaurian as yet known. It first appears in the Oxford Clay, and it is by no means certain that the genus survived even in the period of the Kimmeridge Clay, although possibly the so-called *Ichthyosaurus entheciodon*, in which the teeth are very small, may be a related form.

It is true that a small species of Ichthyosaur from the Cambridge Greensand has also been referred to the same genus under the name *Ophthalmosaurus cantabrigiensis* by Mr. Lydekker, but very little is known of this animal and the presence of facets for three bones on the distal end of the humerus does not seem sufficient evidence, since this character is not confined to *Ophthalmosaurus*. A portion of a lower jaw from the Upper Greensand of Warminster has also been referred to this genus on account of the small size of the teeth, but in this case also the evidence seems insufficient. It is certain that the Cretaceous Ichthyosaurs that are at all well known are not related to *Ophthalmosaurus*, this being shown by the great development of the teeth (in *I. campylodon*, Carter, from the Gault and later) or by the structure of the paddles (in *I. platydactylus*, Broili, from the Lower Greensand of Hanover).

---

* For an exhaustive account of the American species, see C. W. Gilmore, "Osteology of *Baptanodon*," 'Memoirs of the Carnegie Museum,' vol. ii. (1905) p. 77.

The Ichthyosaurs form a singularly homogeneous group, the earliest known forms being already highly specialised for aquatic life, though some traces of a terrestrial ancestry are retained. The first known member of the order is from the lower beds of the Muschelkalk (Middle Trias) of Germany and Switzerland: this was first described by Quenstedt[*] under the name *Ichthyosaurus atavus*, and it has since been discussed in detail by Fraas[†] and Merriam[‡] under the name *Mixosaurus atavus*. Other remains of Ichthyosaurs from the Middle and Upper beds of the Muschelkalk have been found in various European localities, but they are mostly fragmentary. From the Upper Trias of Northern Italy (the Bituminous shales of Besano, in Lombardy) excellently-preserved Ichthyosaurian skeletons are known and have been described by Bassani[§] under the name *I. cornalianus*. Baur[||] subsequently pointed out that this species was more primitive than the later forms in several respects, and proposed to place it in a separate genus *Mixosaurus*. The more important of the characters that seem to point to a terrestrial ancestry are, first, the elongation of the epipodial bones (radius and ulna, tibia and fibula) and their contraction in the middle to form more or less of a shaft; and, second, the differentiation of the teeth, those in the maxillary region being stout and blunt, those in the front of the jaws sharp and conical. In the later forms the epipodials are shortened up and show little or no trace of ever having possessed a shaft, and the teeth are sharp, conical, and numerous, the evolution of the Ichthyosaurs in this last respect, as in some others, showing an interesting case of parallelism with that of the Toothed Whales. A more detailed account of these Italian Ichthyosaurs has been given by Repossi[¶] and by Merriam (*op. cit.*). Remains of Triassic Ichthyosaurs are by no means confined to Europe: numerous forms have been described from Spitzbergen and the United States, and one

---

[*] F. A. Quenstedt, 'Petrefaktenkunde,' 1st ed. (1852) p. 129.

[†] E. Fraas, 'Die Ichthyosaurier der Süddeutschen Trias- und Jura-Ablagerungen' (1891) p. 37.

[‡] J. C. Merriam, "Triassic Ichthyosauria, with Special Reference to the American Forms," Mem. University of California, vol. i. no. 1 (1908) p. 90.

[§] F. Bassani, "Sul fossili e sull' eta degli Schisti bituminosi triasici di Besano," Atti Soc. Ital. Sci. Nat. vol. 29 (1886) p. 20.

[||] G. Baur, "Ueber den Ursprung der Extremitäten der Ichthyopterygia," Berichte ueber der XX. Versam. des Oberrh. geol. Ver. vol. xx. (1887).

[¶] F. Repossi, "Il Mixosauro degli strati triasici di Besano in Lombardia," Atti Soc. Ital. Sci. vol. 41 (1902) p. 361.

species * probably of this age is known from New Zealand, showing that very early in its history the group had become cosmopolitan. Triassic Ichthyosaurs from Spitzbergen were first described by Hulke † in 1873 under the names *Ichthyosaurus nordenskioldi* and *I. polaris*; these species have since been referred by Dames ‡ to *Mixosaurus* and by Yakolew § and Merriam ‖ to the American genera *Cymbospondylus* and *Shastasaurus*. In a paper lately published Wiman ¶ describes a quantity of new material, and finds that while *I. nordenskioldi* is referable to *Mixosaurus*, *I. polaris* is to be placed in a distinct genus, *Pessosaurus*. At the same time he describes a new genus, *Pessopteryx*, including several species. In Wiman's paper (p. 131, fig. 3) there is an interesting restoration of *Mixosaurus*, showing the form of the tail-fin, which in many of these early forms was supported by elongated neural spines, and in some cases chevrons, while the sharp deflexion of the posterior part of the tail found in later forms was only slightly indicated.

The most important series of Triassic Ichthyosaurs is from the United States, where remains referred to several genera have been described in detail by Merriam **. These are from the Middle Trias of Nevada and the Upper Trias of California. The principal genera are *Cymbospondylus*, *Toretocnemus*, *Merriamia* ††, *Delphinosaurus*, *Shastasaurus*, several species of some of these genera being known. Merriam has elaborately tabulated the characters distinguishing these Triassic Ichthyosaurs from the recent types. Some of the more important differences between the Triassic forms and *Ophthalmosaurus* are shown in the following table:—

* J. Hector, "On the Fossil Reptilia of New Zealand." Trans. New Zealand Inst. vol. 6 (1874) p. 355.

† J. W. Hulke, "Memorandum on some Fossil Vertebrate Remains collected by the Swedish Expeditions to Spitzbergen in 1864 and 1868," Bihang k. Svensk. Vet.-Akad. Handl. vol. i. (1873) no. 9.

‡ W. Dames, "Die Ichthyopterygier der Triasformation," Sitzb. Akad. Wiss. Berlin, 1895, p. 1045.

§ N. Yakolew, "Neue Funde von Trias-Sauriern auf Spitzbergen," Verh. Russ.-Kais. Min. Gesell. St. Petersb. ser. 2, vol. 40 (1902) p. 179.

‖ J. C. Merriam, "Triassic Ichthyosauria," Mem. Univ. California, vol. i. no. 1 (1908).

¶ C. Wiman, "Ichthyosaurier aus der Trias Spitzbergens," Bull. Geol. Inst. Upsala, vol. x. (1910) p. 124.

** J. C. Merriam, "Triassic Ichthyosauria, with Special Reference to the American Forms," Mem. Univ. California, vol. i. no. 1 (1908). This memoir contains a very exhaustive account up to the date of its publication of all the known Triassic Ichthyosauria. The only important paper on the subject published since, is that by Wiman referred to above.

†† The name *Merriamia* was substituted for the preoccupied *Leptocheirus* by Boulenger (Proc. Zool. Soc. vol. i. (1904) p. 425). On page 3 of the present volume *Leptocheirus* is employed, the correction not having been made when this part was printed.

Triassic Ichthyosauria.	*Ophthalmosaurus.*
1. The orbits are relatively small and the temporal bar behind them broad.	1. The orbits very large and the temporal bar behind them greatly reduced in width.
2. The maxilla relatively large and the premaxilla correspondingly smaller.	2. The maxilla small and edentulous and the premaxilla relatively very large.
3. Teeth set in distinct sockets and those in the posterior part of the jaws often differing in form from those in front.	3. The teeth, when present, small and loosely fixed in a continuous groove, the anterior and posterior teeth of the same form (as in Jurassic Ichthyosaurs which possess maxillary teeth).
4. The anterior cervical vertebræ separate from one another.	4. The two anterior cervical centra (atlas and axis) fused with one another.
5. The neural spines thick and sometimes circular in section.	5. The neural spines broad and strongly compressed laterally.
6. The zygapophyses of opposite sides separate from one another.	6. The zygapophyses of opposite sides, in most of the vertebræ, in the same plane and united in the middle line.
7. The terminal portion of the caudal series of vertebræ only slightly bent down, the caudal fin being comparatively small and in some cases supported by the elongated neural spines (and sometimes chevrons).	7. The terminal portion of the caudal region of the vertebral column sharply bent down, and, notwithstanding its large size, the caudal fin not supported by the neural spines or chevrons, which are much reduced.
8. Hind limbs larger in proportion to the fore limbs than in the later forms, the reduction of the hind limb, however, already making considerable progress in some species, e. g. *Mixosaurus nordenskioldi*.	8. Fore limbs much larger than the hind limbs.
9. The epipodial bones elongated, with traces of a shaft.	9. The epipodial bones shortened, with no trace of a shaft.
10. The pelvic bones heavy, the ischium and pubis expanded and never fused with one another.	10. Pelvic bones small, the ischium and pubis fused with one another at both ends.

In the Jurassic Ichthyosaurs other than *Ophthalmosaurus*, the differences from the Triassic types above enumerated are, of course, nearly equally well marked; but those numbered 1, 2, 8, and 10 are especially well illustrated by the Oxford Clay type.

As to the origin of *Ophthalmosaurus* there is no certainty, but probably it was derived from one of the "latipinnate" group of Ichthyosaurs. There is nothing in the structure of the skull that is opposed to this suggestion, and the arrangement of the bones in the paddles seems rather to support it. It is unfortunate that no specimens of the paddles have been collected with the bones in an undisturbed

INTRODUCTION.

condition, but so far as the fore paddle is concerned it is believed that the specimen figured on Pl. II. fig. 6 represents as nearly as possible the actual arrangement of the bones, every piece having been carefully numbered and a sketch-plan of their arrangement having been made before their removal from the matrix. The circumstance that the ossicles do not fit together in a close pavement, as in the paddles of *Ichthyosaurus*, but were surrounded by a considerable amount of cartilage, adds to the uncertainty as to the precise arrangement. It can be seen that the intermedium supported two digits, as in the typical latipinnate forms (e. g., *Ichthyosaurus communis, I. intermedius*); but the paddle differs from those of the earlier forms owing to the fact that its width has been still further increased by the great enlargement of the pisiform, which has acquired an articulation with the distal end of the humerus, in some cases almost as large as that possessed by the radius; the row of ossicles supported by the pisiform become enlarged and form a well-developed digit; a preaxial row of small sesamoid ossicles may also be present. There are traces of the widening of the paddle by the increased size of the pisiform and its digit even in the Triassic *Mixosaurus*, and probably some early Jurassic descendant of that genus is the ancestor of *Ophthalmosaurus*, though at present no species is known to which that position can be definitely assigned. There seems to be no reason for the suggestion that *Ophthalmosaurus* is descended from *Shastasaurus*, for in that Triassic genus the digits have already undergone much reduction, there being probably, according to Merriam, only two large digits and one reduced digit in the manus. In the later forms of *Ichthyosaurus* the broadening of the fore paddle which occurs in *Ophthalmosaurus* may be effected in other ways; thus in *Ichthyosaurus extremus*, described by Boulenger * from an unknown locality and horizon, but now known to be almost certainly of Kimmeridgian age, the intermedium is thrust between the radius and ulna, and articulates with the humerus by a well-marked facet, so that that bone comes to resemble closely the humerus of *Ophthalmosaurus*, and if found isolated might be mistaken for it. The width of the paddle in this case is also added to by the presence of a row of sesamoid ossicles on both the preaxial and postaxial borders. Another method of widening is found in *Ichthyosaurus platydactylus*, described by Broili † from the Cretaceous (Aptian) of Hanover; in this species, although it is

---

* Proc. Zool. Soc. vol. i. (1904) p. 424.
† 'Palæontographica,' vol. 54 (1907-8) p. 139, pls. xii. & xiii.

clearly a member of the longipinnate group, the fore paddle attains great width through the addition of at least two rows of supplementary ossicles on the radial side and one on the ulnar side.

Ophthalmosaurus, with its powerful tail-fin, pointed head, and porpoise-like body, must have been a very swift and powerful swimmer, even for an Ichthyosaur, and probably lived in the open sea like most of the Toothed Whales of to-day; like them, too, it was no doubt capable of diving and swimming at considerable depths, the structure of the auditory apparatus, in the opinion of Dollo [*], being specially adapted for use under great pressures such as the animal would be subjected to at some distance beneath the surface. Although in its mode of life *Ophthalmosaurus* probably did not differ greatly from other members of the order, the reduction of the dentition indicates that its food probably differed from theirs, though of its nature nothing is known.

All the Plesiosaurs described in the present volume are members of the Family Elasmosauridæ, characterised especially by the structure of the shoulder-girdle, in which, in the adult, the scapulæ meet in a median symphysis, which is continuous posteriorly with the symphysis of the coracoids. The ingrowth of the scapulæ towards the middle line takes place beneath the clavicular arch, which thus comes to lie on the visceral surface of the ventral rami of the scapulæ, which usurp its functions. The consequence of this is, that the clavicles and interclavicles undergo reduction in varying ways. In some genera all the elements of the clavicular arch persist in a reduced form, in others the clavicles or interclavicle may dwindle away to mere vestiges. These varied conditions of the clavicular arch supply some of the chief characters employed in defining the different genera. If Professor Seeley's [†] restoration of the shoulder-girdle of *Eretmosaurus rugosus* be correct, it would appear that the arrangement of the coracoids and scapulæ found in the Elasmosauridæ had already come into existence in the period of the Lower Lias; and, at any rate, it seems certain that it had done so in the Upper Lias, for Mr. D. M. S. Watson [‡] has lately described from beds of that age at Whitby, a shoulder-girdle of *Plesiosaurus homalospondylus* (referred by him to a new genus *Microcleidus*), in which the form and

---

[*] L. Dollo, "L'audition chez les Ichthyosauriens," Bull. Soc. Belge Géol. etc. vol. xxi. (1907) p. 157.

[†] Quart. Journ. Geol. Soc. vol. xxx. (1874) p. 445.

[‡] Mem. & Proc. Manchester Lit. & Phil. Soc. vol. liv. (1909-10) no. 4, p. 4.

arrangement of all the elements are almost exactly as in *Cryptocleidus*. It should, however, be noted that neither *Eretmosaurus rugosus* nor *Microcleidus homalospondylus* fall within the limits of the family Elasmosauridæ, because their cervical ribs are double-headed, a condition not found in the true Elasmosaurs.

The genus *Elasmosaurus* itself was founded by Cope * for the reception of several species of Plesiosaurians from various Cretaceous deposits in Kansas and Nebraska. The type species was described by Cope from an imperfect skeleton, consisting of the greater part of the vertebral column and the imperfect limb-girdles, the pectoral girdle showing the peculiar structure which has been regarded as of sufficient importance to justify the establishment of a distinct family for the reception of the genera in which it occurs. One peculiarity of *Elasmosaurus* proper is the enormous length of the neck, which consists of no less than 76 vertebræ; in *Murænosaurus*, the English genus in which the neck is longest, there are only 44 cervical vertebræ.

Of the numerous other genera of American Plesiosaurs many are very imperfectly known: some almost certainly belong to this family (e. g., *Cimoliosaurus*, in which the shoulder-girdle is not yet known). In the genera such as *Dolichorhyncops* and *Brachauchenius*, which have been described in detail by Williston †, the shoulder-girdle is not Elasmosaurian; these animals seem to approach rather the Pliosaurian type, though they differ from the typical Pliosauridæ in possessing single-headed cervical ribs, and in some respects the structure of the skull seems to be very different.

In their habits the Plesiosaurs probably differed widely from the Ichthyosaurs. In the first place, their mode of swimming was quite different, propulsion through the water being effected entirely, or almost entirely, by the oar-like paddles, the tail being short and, so far as is known, possessing no fin, or at most a very small one. This manner of swimming and the great length of the neck are characters preventing the supposition that these animals moved at a great pace beneath the water, and it is much more likely that they lived mainly at the surface and at no great distance from the shore. The shorter-necked Pliosaurs are more

---

* Proc. Acad. Nat. Sci. Philadelphia, 1868, p. 68.
† "North-American Plesiosaurs, Pt. I.," Field Columbian Museum—Geology, vol. ii. no. 1 (1903); "North-American Plesiosaurs: *Elasmosaurus, Cimoliosaurus,* and *Polycotylus*," Amer. Journ. Sci. [4] vol. xxi. (1906) p. 221; "The Skull of *Brachauchenius*, &c.," Proc. United States Nat. Mus. vol. xxxii. (1907) p. 477; "North-American Plesiosaurs: *Trinacromerum*," Journ. Geol. vol. xvi. (1908) p. 715.

adapted for a pelagic life, but still to a much smaller degree than the whale-like *Ophthalmosaurus*.

If these animals, as we suppose, lived near the shore, the conditions of life would be much more various than in the case of a truly pelagic animal, and this would probably account for the much greater variety of form found among them than among the Ichthyosaurs. One of the most remarkable circumstances about these Oxford Clay reptiles is the occurrence in a limited area of so large a number of closely related species and genera, in the case both of the Plesiosaurs and of the Crocodiles. This can be reasonably accounted for by supposing that the conditions under which the different forms lived presented considerable variety, some, for instance, living in shallow, some in deeper water, some perhaps in swamps, and some in rivers or river-estuaries. For the same reasons, although the Ichthyosaurs *Ophthalmosaurus* and *Baptanodon* are generically identical, it by no means follows that the American Plesiosaurs contemporary with them will, when better known, be found to be closely similar to the English species.

Although in the Cretaceous period the Plesiosauria had spread over the whole world, being known not only from Europe and North America but also from Asia, South America, South Africa (a species discovered lately), Australia, and New Zealand, it is not certain how far the Elasmosauridæ spread, for in most cases too little is known about the skeleton of these foreign species to make it possible to determine to what family they belong.

The Plesiosaurs were no doubt predaceous, their long sharp teeth being well adapted for the prehension of living prey, which would probably be swallowed whole. The occurrence of numerous stones in the stomach of these animals, first observed by Mr. Thomas Codrington * in a Plesiosaur from the Upper Greensand of Wiltshire, and since noticed in the case of various English and American species, may indicate that the food was broken up in a muscular stomach by the aid of these stones, much as in the gizzard of a bird. No specimen of an Elasmosaur in which the stomach-stones are preserved has been collected from the Oxford Clay, but in the case of a Pliosaur, *Peloneustes*, Mr. Leeds has obtained a hard mass lying within the ribs, containing many stones of various sizes, from

* 'Wiltshire Archæological Magazine,' vol. ix. (1863) p. 170.

that of a small hen's egg downwards, and no doubt representing the fossilized contents of the stomach. The stones are of various kinds, including quartz, sandstone, and gneiss; for the most part they are rather angular with the angles somewhat rounded off. The mass in which the stones are embedded consists mainly of angular grains of quartz-sand of various sizes, and mingled with these are numerous hooks from the arms of Cuttle-fishes, with black masses which show the characteristic structure of portions of the ink-bags of the same creatures. It is notable that the stomach does not seem to contain any of the hard "guards" of the Belemnite shell, so that probably the animal either bit off the soft anterior portion of its prey or perhaps, as Mr. Crick has suggested, fed on some such form as *Geoteuthis*, in which the hard parts were not present.

Although no doubt both the long-necked Elasmosaurs and the short-necked Pliosaurs could catch their prey on the surface, the former probably fed largely on animals living at the bottom, reaching down with their long necks much as do swans. At least, this manner of feeding would account for the tendency to increase the length of the neck in this group, for such increase would be of considerable advantage to the animals in widening their radius of action in the search for food. The longest-necked Plesiosaur at present known is *Elasmosaurus* itself, from the Cretaceous of Kansas: in this reptile, as already noted, there were no less than 76 cervical vertebræ, the total length of the neck being about twenty-three feet, while the body was only about nine feet long. Some such explanation as that suggested above seems necessary to account for the otherwise apparently disproportionate development of this part of the body. The comparative lack of flexibility of the neck, especially of the posterior portion, in some of these reptiles would not be any disadvantage, because the whole body would probably be tilted up much as it is in the case of birds feeding in a similar way.

Any discussion as to the relationship of the Sauropterygia to the other reptiles is deferred till the Pliosaurs have been described.

CHARLES W. ANDREWS.

Department of Geology,
July 1910.

SYSTEMATIC INDEX.

	Page
Order ICHTHYOPTERYGIA	1
Family OPHTHALMOSAURIDÆ	2
Genus Ophthalmosaurus	2
,, icenicus	61
Order SAUROPTERYGIA	77
Suborder *PLESIOSAURIA*	77
Family ELASMOSAURIDÆ	77
Genus Murænosaurus	77
,, leedsi	120
,, durobrivensis	127
,, platyclis	134
Genus Picrocleidus	139
,, beloclis	140
,, sp.	146
Genus Tricleidus	149
,, seeleyi	149
Genus Cryptocleidus	164
,, oxoniensis	164

FRONTISPIECE.

THE photograph represents the nearly complete skeleton of *Cryptoclidus oxoniensis* (R. 2860), as mounted in the Gallery of Fossil Reptiles. All the bones belong to a single individual and, with the exception of two or three ribs and neural spines, no restoration has been made on the side shown. The total length of the skeleton is 11 feet.

LIST OF ILLUSTRATIONS IN THE TEXT.

Fig.			Page
1.	*Ophthalmosaurus*:	basioccipital and basisphenoid	5
2.	,,	exoccipital and supraoccipital	7
3.	,,	pro-otic, opisthotic, and stapes	10
4.	,,	reconstruction of the posterior region of the skull	12
5.	,,	basisphenoid	13
6.	*Ophthalmosaurus* and *Ichthyosaurus*: basisphenoid		14
7.	*Ophthalmosaurus*:	basisphenoid and parasphenoid	15
8.	,,	squamosal and postorbital	17
9.	,,	quadrate	18
10.	,,	quadrato-jugal and jugal	20
11.	,,	lachrymal	21
12.	,,	nasal	22
13.	,,	maxilla	23
14.	,,	parietal	25
15.	,,	part of upper region of skull	26
16.	,,	postfrontal	27
17.	,,	pterygoid	29
18.	,,	palatine (?)	29
19.	,,	vomer	30
20.	,,	posterior portion of right ramus of mandible	32
21.	,,	surangular	33
22.	,,	articular	34
23.	,,	restoration of skull and mandible	35
24.	,,	centra of axis and atlas vertebræ	37
25.	,,	centra of anterior cervical vertebræ	39
26.	,,	dorsal and caudal vertebræ	40
27.	,,	caudal vertebræ	41
28.	,,	caudal vertebræ	42
29.	,,	left half of neural arch of anterior cervical vertebræ	43
30.	,,	neural arch of a dorsal vertebra	44
31.	,,	sections of vertebral centra and the upper end of a dorsal rib	45
32.	,,	coracoid	47
33.	,,	scapula	48

Fig.			Page
34.	*Ophthalmosaurus*:	clavicle and interclavicle	50
35.	,,	restoration of shoulder-girdle	51
36.	,,	humerus	52
37.	,,	fore paddle	55
38.	,,	ilium	57
39.	,,	ischio-pubis	58
40.	,,	ischio-pubis	59
41.	,,	hind paddle	60
42.	,,	restoration of skeleton	62
43.	*Murænosaurus*:	basioccipital	79
44.	,,	basioccipital, basisphenoid, exoccipital, and parasphenoid	80
45.	,,	supraoccipital and exoccipital-opisthotic	83
46.	,,	restoration of skull	85
47.	,,	restoration of palate	88
48.	,,	posterior portion of mandibular ramus	90
49.	,,	atlas and axis vertebræ	92
50.	,,	anterior cervical vertebræ	94
51.	,,	posterior cervical vertebræ	95
52.	,,	pectoral and first dorsal vertebræ	96
53.	,,	anterior dorsal vertebra	98
54.	,,	middle dorsal vertebra	99
55.	,,	posterior dorsal vertebra	100
56.	,,	sacral vertebra and rib	101
57.	,,	middle caudal vertebra	102
58.	,,	posterior caudal vertebræ	103
59.	,,	caudal vertebræ and chevrons	104
60.	,,	ventral and dorsal ribs	106
61.	*Plesiosaurus* and *Nothosaurus*: shoulder-girdle		108
62.	*Murænosaurus*: shoulder-girdle		109
63.	,,	fore and hind paddles	112
64.	*Nothosaurus*: pelvis		114
65.	*Murænosaurus*: pelvis		116
66.	,,	restoration of skeleton	118
67.	,,	*durobrivensis*: shoulder-girdle	128
68.	,,	*platyclis*: shoulder-girdle	134
69.	,,	,, humerus and femur	135
70.	*Picrocleidus*: shoulder-girdle		140
71.	,,	sacral vertebræ and ribs	148
72.	*Tricleidus*: exoccipital-opisthotic		150
73.	,,	basioccipital, basisphenoid, and parasphenoid	151
74.	,,	basis cranii and pterygoids	153
75.	,,	left squamosal and quadrate	155
76.	,,	shoulder-girdle	158
77.	,,	left fore paddle	160

LIST OF ILLUSTRATIONS IN THE TEXT.

Fig.			Page
78.	*Cryptocleidus* :	atlas and axis	169
79.	,,	anterior cervical vertebræ	170
80.	,,	posterior cervical vertebra	171
81.	,,	middle dorsal vertebra	171
82.	,,	posterior dorsal vertebra	172
83.	,,	sacral vertebra and ribs	173
84.	,,	anterior caudal vertebra	173
85.	,,	posterior caudal vertebræ	174
86.	,,	plastron of ventral ribs	175
87.	,,	shoulder-girdle	176
88.	,,	shoulder-girdle	179
89.	,,	immature shoulder-girdles	180
90.	,,	portions of fore paddles	182
91.	,,	portion of fore and hind paddles	184
92.	,,	immature pelvis	186
93.	,,	hind paddle	187
94.	,,	restoration of skeleton	188

A DESCRIPTIVE CATALOGUE

OF THE

MARINE REPTILES

OF

THE OXFORD CLAY.

Order ICHTHYOPTERYGIA.

CARNIVOROUS marine reptiles with a cetacean-like body and no visible neck; a dorsal fin without skeletal supports, and a large vertical caudal fin, of which the lower lobe is strengthened by the downwardly-turned end of the vertebral column. A plastron of ventral ribs present, but no dermal armour.

Skull large, with a more or less elongated rostrum composed mainly of the premaxillæ. External nares a little in front of the very large orbits, in which there is always a ring of ossified sclerotic plates. A large parietal foramen. Supratemporal fossa large, post-temporal fossa small, and lateral temporal fossa absent; a vacuity enclosed between the quadrate and quadrato-jugal. Postorbital and postfrontal bones distinct; lachrymal and prefrontal also distinct. Quadrate immovable, being closely united below with the pterygoid, above with the squamosal and supratemporal, which remain separate. In the otic region, pro-otic and opisthotic elements distinct; stapes (where known) short and stout, extending from basioccipital to quadrate. In the palate, pterygoids extending forwards to meet the vomers and excluding the palatines from the middle line; an epipterygoid (*columella cranii*) present; a large parasphenoid extending forwards between the pterygoids; apparently no transverse bone. Teeth simply conical, usually with a vertical Labyrinthodont-like folding of their walls; restricted to the edge of the jaws; heterodont and set in distinct sockets in some early genera (e. g. *Mixosaurus*); homodont and set in continuous dental grooves in the typical and later genera.

Vertebral centra amphicœlous and usually very short. A varying number of the anterior ribs with a double articulation with the vertebræ. No sacrum.

In the shoulder-girdle, distinct clavicles and an interclavicle, which is usually T-shaped (triangular in *Leptocheirus*, Merriam [*]). Pelvis, at least in the later forms, much reduced, and the ilium not united with the vertebral column. Limbs modified to form paddles, and the fore limb, as a rule, considerably the larger (not in the Triassic genus *Toretocnemus*, Merriam [†]); digits varying in number from three to six and consisting of numerous phalanges.

Range from Trias to Lower Chalk.

Family OPHTHALMOSAURIDÆ.

Specialised Ichthyosaurs in which the orbit is extremely large and the dentition reduced in the adult to a number of small teeth, which are loosely set in the anterior half of the jaws only. The humerus articulates distally with three bones, and the fore paddle is much larger than the posterior, which is greatly reduced; the pubes and ischium are fused together.

Middle and Upper Jurassic of Europe and North America, with doubtful representatives in the Upper Greensand of England.

Genus OPHTHALMOSAURUS, Seeley.

[Quart. Journ. Geol. Soc. vol. xxx. (1874) p. 699.]

1879. *Sauranodon*, O. C. Marsh, Amer. Journ. Sci. [3] vol. xvii. p. 85. (Name previously employed by Jourdan for a Rhynchocephalian [‡].)
1880. *Baptanodon*, O. C. Marsh, Amer. Journ. Sci. [3] vol. xix. p. 491.
1902. *Microdontosaurus*, C. W. Gilmore, Science, n. s. vol. xvi. p. 914.

Orbit very large; teeth small, confined to the anterior half of the jaws, at least in the adult. Clavicles uniting in a complex suture (sometimes fused) and embracing the anterior bar of the interclavicle; coracoid normally without posterior notch. Humerus with strong trochanteric ridge and articulating distally with three elements. Pelvis greatly reduced; ischium and pubis normally fused with one another at the ends and enclosing between them a small foramen. Hind limb very small, femur articulating distally with two elements only.

Middle and Upper Jurassic.

This genus was established (*loc. cit. supra*) in 1874 by Prof. H. G. Seeley for the

---

[*] "New Ichthyosauria from the Upper Triassic of California," Bull. Geol. Dept. Univ. California, vol. iii. (1903) p. 253.
[†] *Tom. cit.* p. 259.
[‡] Quoted by Gervais in Comptes Rendus Ac. Sci. Paris, vol. lxxiii. (1871) p. 605, from Jourdan's MS.

reception of the species *O. icenicus*. The type specimen was a shoulder-girdle from the Oxford Clay of Peterborough: with this there were also associated other parts of the skeleton of the same individual, including portions of the skull, mandible, numerous vertebræ, portions of ribs and neural arches, and some paddle-bones (see Catalogue, p. 63, R. 2133). The feature regarded as specially characterising the genus was the union of the clavicles by suture in the middle line so as closely to embrace the anterior bar of the interclavicle. In the same paper a fore limb of another individual was described and its chief peculiarity, viz. the articulation of the distal end of the humerus with three elements, pointed out. It is unfortunate that the type shoulder-girdle is greatly diseased and deformed, the right coracoid being an almost shapeless mass of bone, while the left has a deep posterior notch which is entirely wanting in all normal specimens. The structure of the shoulder-girdle was further discussed by Prof. Seeley in his paper " Further Observations on the Shoulder-Girdle and Clavicular Arch in the Ichthyosauria and Sauropterygia" *. Additional information concerning the genus has been given by Lydekker †, A. S. Woodward ‡, and Bauer §. The last-mentioned writer has described the occipital and otic regions of the skull from material in the collection now under discussion, but, as will be shown below, his account of the arrangement of the otic elements is not quite accurate.

The Ichthyosaur from the Upper Jurassic beds of the United States, originally described by Marsh under the name *Sauranodon* ‖, afterwards emended to *Baptanodon* ¶, has lately been the subject of several important papers by Knight **, and especially by Gilmore. Knight describes an imperfect skeleton and gives a number of reasons for regarding *Baptanodon* as distinct from *Ophthalmosaurus*, with which several writers have considered it as identical. Gilmore first recorded the occurrence of teeth in *Baptanodon* ††, but in his first note he referred the specimen in which teeth were observed to a new genus *Microdontosaurus*, a name subsequently withdrawn. He has lately published the most detailed account ‡‡ of the skeleton of this form that has yet appeared, and while pointing out that many of the differences between *Baptanodon* and *Ophthalmosaurus* referred to by Knight, do not really exist, and were partly the consequence of the bad state of preservation of the specimens

---

* Proc. Roy. Soc. vol. liv. (1893) p. 149.
† Catal. Foss. Rept. Brit. Mus. pt. ii. (1889) p. 8; also pt. iv. (1890) p. 268.
‡ 'Vertebrate Palæontology' (1898) p. 183.
§ " Osteologische Notizen über Ichthyosaurier," Anat. Anzeiger, vol. xviii. (1900) p. 581.
‖ " A new Order of Extinct Reptiles (Sauranodonta) from the Jurassic Formation of the Rocky Mountains," Amer. Journ. Sci. [3] vol. xvii. (1879) p. 85. Also " The Limbs of Sauranodon, with a Notice of a new Species," *loc. cit.* [3] vol. xix. (1880) p. 169.
¶ Amer. Journ. Sci. [3] vol. xix. (1880) p. 491.
** " Notes on the Genus *Baptanodon*," Amer. Journ. Sci. [4] vol. xvi. (1903) p. 76.
†† Science, n. s. vol. xvi. (1902) p. 913, and vol. xvii. (1903) p. 750.
‡‡ " Osteology of *Baptanodon*," Mem. Carnegie Museum, vol. ii. (1905) p. 77; also *tom. cit.* p. 325.

examined, nevertheless he still considers that the American and English forms are generically different. Apparently the only reasons for this belief worth considering are: (1) that in *Baptanodon* the clavicles are fused in the middle line, instead of merely uniting in a close suture; (2) in the American forms the anterior cervical vertebræ are uniformly biconcave at the ends of their centra, while in the English types the author states that only the middle portion of the centrum is cupped, the concave portion being surrounded by a flattened area; (3) *Baptanodon* is said to have an additional digit in the fore limb. With regard to these differences, taking them in reverse order, it may be said that the number of digits is by no means certain, and, in fact, Knight's figure of the fore paddle of *Baptanodon*, apparently the only one known in which the bones are *in situ*, lends no support to the view that six digits were present; and even if it did so, it is by no means impossible that an extra row of phalanges may not occasionally have been present in *Ophthalmosaurus*. As to the form of the vertebræ, it may be said that in many cases the anterior cervicals of *Ophthalmosaurus* are biconcave without any broad flattened area round the concavity. Finally, with regard to the fusion of the clavicles in *Baptanodon*, it would be somewhat remarkable if two bones so closely interlocking as the clavicles of *Ophthalmosaurus* did not, at least sometimes, fuse in old age, and, as a matter of fact, this seems to have actually happened in some specimens; in any case, the character does not appear to be of generic value. It seems, therefore, that the English and American species may be regarded as belonging to a single genus, which must be called *Ophthalmosaurus*, that name having the priority, a conclusion already arrived at by E. Fraas [*] and other writers. The case for this identity is further supported by the fact that the associated invertebrate fauna proves that the beds in which the remains occur were contemporary, and also that there are found in the American deposits remains of a Plesiosaur called by Marsh [†] *Pantosaurus* and clearly identical with *Murænosaurus* of the Oxford Clay of England. If further proof of the identity of the American and English genera is needed, it will be found in comparing the present account of the skeleton of the latter with the excellent and detailed account of the former given by Mr. C. W. Gilmore in the papers referred to above.

Skull.—The skull is represented in the collection by a number of more or less nearly complete examples. The description of the bones of the back of the skull is founded mainly on a series of separate and uncrushed bones of a very large individual (R. 2162), while the account of the facial region is taken from a smaller skull (R. 2180), in which nearly all the bones are preserved separate from one another and only slightly crushed. Only a few bones of the skull of the type specimen are preserved (Pl. I. figs. 11–15), and to these reference will be made below.

[*] "Weitere Beiträge zur Fauna des Jura von Nordost-Groenland," Meddelelser om Grønland, vol. xxix. (1904) pt. i. p. 283.

[†] "The Reptilia of the *Baptanodon* Beds," Amer. Journ. Sci. [3] vol. l. (1895) p. 406.

The description of the form and relations of the bones of the occipital and auditory regions of the skull is rendered difficult by the circumstance that a considerable amount of cartilage persisted throughout the animal's life. The consequence of this is, that not only does the form of the several elements vary considerably according to the degree to which ossification has proceeded, but also, owing to the persistence of the cartilage, actual surfaces of contact between the different bones are only found in a

Text-fig. 1.

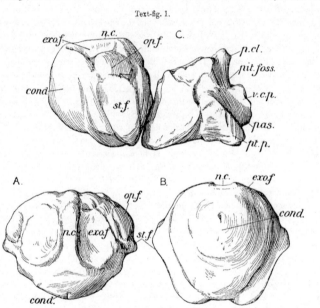

Basioccipital and basisphenoid of *Ophthalmosaurus*: A, basioccipital from above; B, basioccipital from behind; C, basioccipital and basisphenoid from side. (R. 2162, ⅔ nat. size.)

cond., occipital condyle; *exo.f.*, facet for exoccipital; *n.c.*, neural canal; *op.f.*, facet for opisthotic; *pas.*, posterior part of parasphenoid; *p.cl.*, posterior clinoid processes; *pit.foss.*, pituitary fossa; *pt.p.*, pterygoid processes; *st.f.*, facet for stapes; *v.c.p.*, lower cylindrical processes of basisphenoid.

few cases, and usually the separate elements have been displaced and scattered, so that their original position is difficult to make out. The separate elements are first described and then a restoration of this part of the skull is attempted.

The *basioccipital* (Pl. I. figs. 13, 14; text-figs. 1 & 4) is a short and very massive bone. It forms the whole of the occipital condyle (*cond.*), which is sessile and forms less than a hemisphere. The condyle is usually about equally convex in all directions, but sometimes (*e. g.* in type specimen, Pl. I. figs. 13, 14) it may be somewhat pinched in laterally towards its upper end; the outline is nearly circular, but it is flattened above for a short distance (*n.c.*), where it forms the lower border of the foramen magnum; its surface is usually marked by a series of slight concentric ridges, and there is near its middle a small pit or dimple, probably marking the original position of the notochord. The upper surface of the bone (text-fig. 1, A) is occupied in the middle line by a smooth, slightly concave surface (*n.c.*), extending from the upper border of the condyle to the anterior edge; this surface, which is the floor of the neural canal, is narrowed somewhat in the middle by the encroachment of the large roughened concave surfaces for union with the exoccipitals (*exo.f.*). In front of, and a little to the outer side of these surfaces there is in many specimens a slight prominence terminating in a smooth facet (*op.f.*), which appears to have supported the anterior portion of the opisthotic. On the sides of the bone in front of the condyle there is a smooth area slightly concave from before backwards, and in front of this a broad roughened surface (*st.f.*) looking outwards and a little downwards, with which the head of the stapes articulates. The ventral surface is also occupied by a smooth area which ends in front in a straight or slightly concave border, along which the bone is in contact with the basisphenoid. The anterior face slopes somewhat backwards and is entirely occupied by a coarsely roughened surface for cartilage, usually divided into two bosses by a slight median groove; it is clear that even in old individuals the basioccipital and basisphenoid were only in contact at most along their ventral edge, and were separated above by a thick wedge of cartilage (text-fig. 1, C).

The *exoccipitals* (*exo.*, text-fig. 2, A & B, also text-fig. 4) are short, stout, columnar bones which form the lower part of the lateral border of the *foramen magnum*. At their ventral end (*boc.f.*) they are considerably expanded, their base extending forwards in a long tongue-like process, and in consequence of this their surface for union with the basioccipital is very extensive. Their flattened posterior face seems to have sloped somewhat forwards; near the middle of its outer border it is perforated by a large foramen (XII'), the inner opening of which lies at about the middle of the inner (cranial) surface. This is concave from above downwards, and in addition to the large foramen just referred to, there is an oblique slit-like opening (XII) just anterior to it. The anterior border is strongly notched, the notch apparently forming the posterior border of the so-called *foramen jugularis* (*j.for.*). Judging from the arrangement of the nerve-exits of this part of the skull of *Hatteria*, as described by Osawa[*], it seems probable that the two foramina (XII, XII') perforating the exoccipital, transmitted two branches of the XII nerve, of which the posterior is the larger, and that the IX–XI nerves passed out through the jugular foramen, the hinder border of which is formed by the exoccipital as above described. The outer face of the

---

[*] Archiv f. mikroscop. Anatomie, vol. li. (1897) pp. 494–5.

exoccipital is short and strongly concave from above downwards—in fact, forming merely a deep groove, in the middle of which is the outer opening of the anterior hypoglossal foramen (XII) above noticed, while anteriorly it passes into the border of the jugular notch. The upper end bears two facets, one roughly triangular and slightly convex for union with the supraoccipital; this surface looks directly upwards. The other surface looks outwards, upwards, and forwards, while above it is continuous with the supraoccipital surface; this facet, in some cases at least, was in contact with a corresponding surface on the upper posterior angle of the opisthotic (q. v.), but in other specimens the two elements were probably separated by a pad of cartilage.

Text-fig. 2.

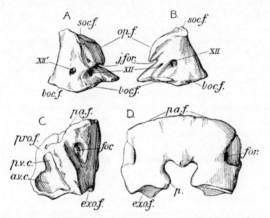

Exoccipital and supraoccipital of *Ophthalmosaurus*: A, right exoccipital from outer side; B, ditto from inner side; C, supraoccipital from side; D, ditto from behind. (R. 2162, ⅔ nat. size.)

a.v.c., depression for anterior vertical semicircular canal; *boc.f.*, facet for basioccipital; *exo.f.*, facet for exoccipital; *for.*, foramen of supraoccipital; *j.for.*, jugular foramen; *op.f.*, facet for the opisthotic; *p.*, process of supraoccipital projecting into foramen magnum; *pa.f.*, facet for the parietal; *pro.f.*, facet for the prootic; *p.v.c.*, depression for the posterior vertical semicircular canal; *soc.f.*, facet for supraoccipital; XII, XII', foramina for the exit of the hypoglossal nerve.

The *supraoccipital* (text-fig. 2, C & D, also text-fig. 4) is a large ∩-shaped bone forming the upper portion of the boundary of the foramen magnum, which is greatly narrowed in this region; the posterior surface of the bone, as a whole, is gently convex from side to side. The form of the edge bordering on the foramen magnum differs in different examples; in some it is simply ∩-shaped, in others (*e. g.* that figured in text-fig. 2, D) there is a median process projecting down from the middle of the upper edge. The opening is, moreover, usually more or less constricted by a pair of blunt processes (*p.*) situated near the lower end of the opening, and it seems most likely that the actual neural canal was only that portion of the opening below these processes. The lower ends of the arch are occupied by gently concave triangular surfaces (*exo.f.*) for union with the exoccipitals. At the upper outer angle of the occipital surface there is a funnel-shaped depression, at the bottom of which is a large foramen (*for.*), which perforates the bone, passing into the cranial cavity by a large smoothly rounded aperture. The function of this opening, which does not seem to have been observed elsewhere, is doubtful; possibly it transmitted a large blood-vessel. From its position it seems not unlikely that this opening may mark the line of separation between a primitively separate epiotic and the true supraoccipital; for, although Baur [*] has stated that no trace of a separate epiotic element has ever been observed in Reptilia, this seems to apply only to skulls in a comparatively advanced state of ossification, for a separate epiotic has been figured by Parker [†] in young embryos of several types (e.g. *Tropidonotus* and *Lacerta*). Another possible explanation of this opening is, that it may have given passage to a part of the enlarged upper end of the *ductus endolymphaticus* of the ear, such as seems to occur in some Geckoes, in which, according to Wiedersheim [‡], a portion of the enlarged *saccus endolymphaticus* lies on the outside of the skull on and among the muscles of the neck, this external portion communicating with that inside the cranium by a duct passing through a foramen between the parietal and auditory capsule (or between the parietal and supraoccipital), which, it is suggested, may be equivalent to the opening here noticed. Wiedersheim believes that the object of this enlarged *saccus*, which is more or less full of fine crystals, is to increase the sensitiveness to sound-vibrations [§]. If the conjecture that some such structure existed in the Ichthyosaurs is correct, it may be supposed to have compensated for the loss of sensitiveness to sound-vibration that must have resulted from the peculiar modification of the stapes, both as to its form, size, and relations to the surrounding bones.

The lateral (epiotic) region of the supraoccipital projects forwards at right angles to the occipital portion of the bone; it is irregularly triangular in outline and bears on its outer surface a smoothly rounded triradiate depression, which marks the position of the inner wall of the anterior and posterior vertical semicircular canals (*a.v.c.*, *p.v.c.*). These are bordered by a roughened edge of varying width for cartilage, the posterior edge being the broadest, the anterior of moderate width, the ventral very narrow and even wanting anteriorly and posteriorly where the channels for the canals reach the

[*] Zool. Anzeig. vol. xii. (1889) pp. 46–47.
[†] Phil. Trans. 1879 (pt. ii.), p. 595, pl. 41. fig. 5; also 1878 (pt. ii.), p. 385, pl. 31. figs. 3, 4.
[‡] Morphologisches Jahrbuch, vol. i. (1876) pp. 495–534, pls. xvii.–xix.
[§] Mr. Boulenger has pointed out to me that the size of this lime-containing sac varies greatly even in different individuals of the same species, and it is often apparently absent; this would, of course, be against regarding this organ as having important auditory functions.

edge of the bone. Probably there was no actual contact either with the prootic or opisthotic, broad tracts of cartilage having intervened between the several otic elements. The upper border of the supraoccipital is gently convex from side to side and is also curved forwards laterally; the edge is broad and occupied by a deep roughened groove (*pa.f.*), indicating that probably there was a pad of cartilage between this bone and the overlying parietal.

The *prootic* (text-fig. 3, A, B) is a very small bone, oval in outline, with the outer surface gently convex in all directions, the convexity being most marked at one end of the long axis. Its inner face bears a triradiate smooth channel corresponding to the anterior vertical and horizontal semicircular canals; round these is a roughened border for cartilage, varying in width, the widest being at the most convex end of the long axis. The precise position of this bone cannot be made out, as it does not appear to have been in actual contact with either of the other otic elements, but was surrounded by cartilage. Probably its position has been most nearly determined by Bauer [*] in his figures given in his paper on the otic bones of this genus.

The *opisthotic* (Pl. I. fig. 15; text-fig. 3, E, F) is a large and solidly constructed bone, the position of which with regard to the surrounding elements is not easy to determine. The following account is founded on an examination of several sets of bones of the occipital region, and may be taken as representing fairly exactly the actual condition of things. It will be seen that the position here ascribed to the opisthotic approaches most nearly to that described by Gilmore and shown in his figure of the skull of *Baptanodon* [†], but differs in the relations with the exoccipitals; on the other hand, the descriptions given by Bauer in *Ophthalmosaurus* [‡], and by E. Fraas [§], Cope [||], and Owen [¶] in *Ichthyosaurus*, differ considerably from the present account. The bone consists of an inner greatly thickened region and an outer short stout process directed upwards and outwards, and terminating in a convex facet (*sq.f.*), which fits closely into a corresponding surface on the inner face of the squamosal, in the angle between the process of that bone which joins the parietals and the downwardly-directed portion which embraces the upper end of the quadrate (see text-figs. 4 & 8, A, B). This outwardly directed bar of the bone bears both on its anterior and still more markedly on its posterior face strong ridges and roughened tuberosities (*t.*) for union with muscles or tendons.

The expanded inner end of the bone is roughly trihedral; ventrally it bears a pair of rather narrow, roughly triangular surfaces (*st.f.* & *st.f'.*) (the posterior of

---

[*] Anat. Anzeig. vol. xviii. (1900) pp. 586–7, figs. 17, 18.
[†] Mem. Carnegie Museum, vol. ii. (1905) p. 85, pl. xi. fig. 2.
[‡] Anat. Anzeiger, vol. xviii. (1900) p. 581.
[§] 'Die Ichthyosaurier der Süddeutschen Trias- und Jura-Ablagerungen,' pl. ii. fig. 3.
[||] Proc. Amer. Assoc. vol. xix. (1870) p. 199, fig. 2.
[¶] 'Foss. Rept. Lias. Form.' pt. iii. (Mon. Pal. Soc. 1881) pl. xxvi. fig. 1.

which is concave, the anterior convex) for union with corresponding facets on the upper surface of the stapes. In both elements these facets are separated by a deep groove (*g.*), so that when the two bones are in apposition a channel is enclosed

Text-fig. 3.

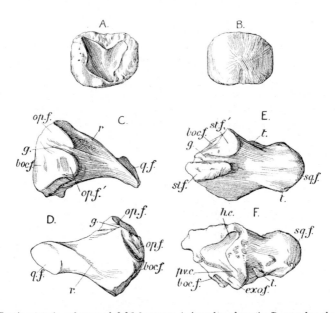

Prootic, opisthotic, and stapes of *Ophthalmosaurus*: A, inner face of prootic; B, outer face of prootic; C, right stapes from above; D, ditto from front; E, right opisthotic from below; F, ditto from above. (A, B, R. 2161; C–F, R. 2162: all ⅔ nat. size.)

boc.f., facet for basioccipital; *exo.f.*, facet for exoccipital; *g.*, grooves in stapes and opisthotic, forming a foramen when the two are articulated with one another; *h.c.*, channel for horizontal semicircular canal; *op.f.*, *op.f'.*, the two facets on the stapes for the opisthotic; *p.v.c.*, channel for posterior vertical semicircular canal; *q.f.*, facet for quadrate; *r.*, ridge on shaft of stapes; *sq.f.*, facet for squamosal; *st.f.*, *st.f'.*, the two facets of the opisthotic for the stapes; *t.*, tuberosities on shaft of opisthotic.

between them, running from within outwards and probably transmitting one of the nerves (? the ninth) issuing from the jugular foramen. Above the posterior of the two facets just referred to, there is a large, roughly semicircular surface (*boc.f.*), by which the bone unites with the basioccipital. Above this again, and separated from it by a notch, is a pointed process, bearing in some cases a small facet (*exo.f.*) for articulation with the outer upper facet of the exoccipital; in cases where the facet was wanting, no doubt the actual junction of the two bones was prevented by the intervention of a pad of cartilage. The junction of the opisthotic with the exoccipital above and the basioccipital below, encloses between it and the former of these two bones an oval vacuity, through which must have run some of the nerves and vessels passing into the jugular foramen which, as above described, notches the front of the exoccipital. The upper and anterior face of the inner end of the opisthotic formed the posterior, outer and lower portion of the auditory capsule, and is deeply impressed by two grooves meeting towards the inner end and marking the position of part of the posterior vertical (*p.v.c.*) and of the horizontal (*h.c.*) semicircular canals. Round these the bone forms a roughened border of varying width for cartilage, and probably it had no contact with either the supraoccipital or epiotic elements. The chief peculiarities of this bone (*e. g.*, its extensive union with the basioccipital and stapes) all seem to tend to increasing the rigidity of the squamoso-quadrate complex.

The *stapes* (Pl. I. fig. 12; text-figs. 3, C, D, & 4), as in other Ichthyosauria, has undergone a most peculiar modification. Instead of being, as in most Reptiles, a slender rod of bone connecting the tympanic drum with the inner ear, it seems to have lost its auditory function and has become a stout bar of bone, lying between the basioccipital and quadrate and helping to add to the rigidity of the latter. The proximal end is greatly enlarged and forms a massive head, the inner face of which is occupied by a surface (*boc.f.*), slightly convex from above downwards, by which it articulates with the corresponding facet on the basioccipital. Above this surface on the upper side of the head are two obliquely elongated facets (*op.f.* & *op.f'.*) separated by a groove; these, as above described, articulate with the corresponding surfaces of the opisthotic, the groove forming the lower half of the channel (*g.*) running between the two elements. The posterior face of the head is flattened and a little roughened; the anterior face is produced ventrally into a blunt triangular prominence, from which a ridge (*r.*) runs obliquely upwards on to the narrow outer process of the bone. This latter terminates in a blunt point, which bears a facet, looking upwards and forwards, for union with the corresponding pit on the inner face of the quadrate. In addition to its inner and outer unions with the basioccipital and quadrate respectively, the lower angle of this bone is wedged into the angle formed by a ventral shelf-like process of the portion of the pterygoid which is united to the inner face of the quadrate (see *pterygoid* below, also text-fig. 4). This union added still further to the rigidity of the occiput.

The arrangement of the various bones described above is figured in text-fig. 4. In this the form of the foramen magnum is well shown, as also is the remarkable manner in which the quadrate is supported by the stapes, opisthotic (indirectly), squamosal, and pterygoid.

The *basisphenoid* (text-figs. 1, 5, 6, 7) is a stout and very massive bone. Its posterior and most of its upper surfaces are greatly roughened and were obviously thickly covered by cartilage in life. In this region the bone is divided into two prominent convex bosses, separated by a deep median groove running from the middle of the upper anterior edge to the lower posterior border. The posterior faces of these prominences are slightly flattened, or even slightly concave, and were directed towards

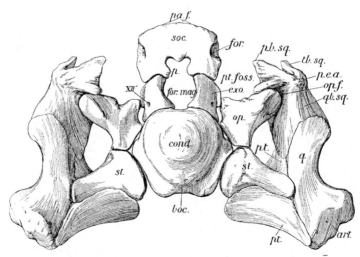

Reconstruction of the posterior region of skull of *Ophthalmosaurus* from behind. (About ½ nat. size.)
art., articular surface of quadrate; *boc.*, basioccipital; *cond.*, occipital condyle; *exo.*, exoccipital; *for.*, foramen in supraoccipital; *for.mag.*, foramen magnum; *op.*, opisthotic; *op.f.*, facet for opisthotic; *p.*, process of supraoccipital projecting into foramen magnum; *pa.f.*, facet for parietal; *p.b.sq.*, parietal branch of the squamosal; *p.e.a.*, postero-external angle of the squamosal; *pt.*, pterygoid; *pt.foss*, post-temporal fossa; *q.*, quadrate; *q.b.sq.*, quadrate branch of the squamosal; *soc.*, supraoccipital; *st.*, stapes; *t.b.sq.*, temporal branch of squamosal; XII', foramen for posterior branch of the hypoglossal nerve.

the corresponding anterior surfaces of the basioccipital, though probably separated from them by a thick pad of cartilage. The presence of the deep median groove above noticed may be a trace of the original ossification of this bone from two lateral centres. The anterior border of the upper surface is raised into a pair of blunt processes separated by a slight notch: these are the posterior clinoid processes (*p.cl.*). Beneath them the anterior face of the bone is at first vertical and then slopes slightly backwards, forming the posterior wall of the very large internal carotid foramen (*i.c.f.*), the anterior wall of which is formed by the upper surface of two prominent bosses, separated by a deep notch and each terminating in a surface for cartilage: these are the processes called by Siebenrock *, in his account of the skull of *Hatteria*, the lower cylindrical processes (*v.c.p.*); they mark the beginning of the presphenoidal region of the *basis cranii*, which remained unossified. Beneath them is the posterior end of the parasphenoid (*pas.*), which is adherent to the ventral face of this bone, as described below. The hollow bounded by the upper surface of the lower cylindrical processes below, and by the posterior clinoid process behind, is the pituitary fossa (*pit.fos.*), into which, as already noted, the very wide carotid canal opens and passes downwards

Text-fig. 5.

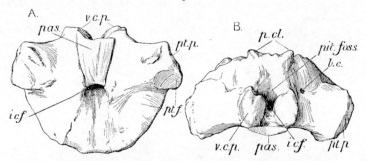

Basisphenoid of *Ophthalmosaurus*: A, from below; B, from the front. (R. 2162, ⅔ nat. size.)

b.c., foramen for a branch of the carotid artery; *i.c.f.*, internal carotid foramen; *pas.*, parasphenoid; *p.cl.*, posterior clinoid processes; *pit.foss.*, pituitary fossa; *pt.f.*, facet for pterygoid; *pt.p.*, pterygoid processes; *v.c.p.*, lower cylindrical processes of basisphenoid.

and backwards, its single ventral aperture being situated about the middle of the ventral face of the bone: this lower opening is usually more or less asymmetrical in form.

In some of the Liassic Ichthyosaurs, as pointed out by Cuvier † and more fully by Maggi ‡, there are two posterior openings, separated, in some cases at least, by the posterior end of the parasphenoid (see text-fig. 6, C); this condition seems to be the most primitive, since in *Hatteria* the paired openings are situated on either side of the posterior end of the parasphenoid. In *Ophthalmosaurus* the condition of the posterior end of the parasphenoid (see below) is very different in different individuals

* " Zur Osteologie des Hatteria-Kopfes," Sitzungsber. k. Akad. Wiss. Wien., math.-naturw. Cl., vol. cii. (1893) p. 250; also translated in Ann. Mag. Nat. Hist. [6] vol. xiii. (1894) p. 297.

† ' Ossements fossiles,' vol. v. pt. 2 (1824) p. 460, pl. xxix. figs. 12 & 13.

‡ " Il Canale Craneo-faringeo negli Ittiosauri etc.," Rendiconti R. Istit. Lombardo, [2] vol. xxxi. (1898) p. 761.

(see text-figs. 5, A; 6, A, B), but in no case does it divide the opening of the carotid canal.

On either side of the pituitary fossa the basisphenoid is produced outwards into the two prominent pterygoid processes (*pt.p.*), the upper surfaces of which look upwards and forwards; at their base there is a small perforation (*b.c.*) leading into the carotid canal and no doubt transmitting a small branch of the internal carotid. The outer ends of the processes are truncated by the facets for the pterygoids looking almost directly outwards and sloping downwards from behind forwards (see fig. 1, C). The ventral surface of the pterygoid processes is roughened for union with the inner plate of the pterygoids, which overlapped and closely united with this part of the ventral face of the basisphenoid, extending back even to the posterior portion of the body, as is shown by the presence of roughened surfaces for union.

When the basisphenoid is looked at from below (text-fig. 5, A) its posterior border is seen to be nearly a semicircle, while the anterior border is nearly straight, except for the slight projections formed by the ends of the lower cylindrical processes (*v.c.p.*)

Text-fig. 6.

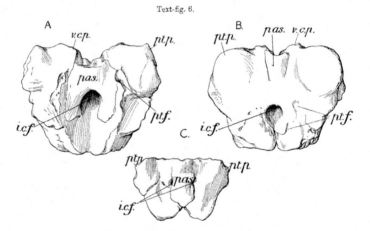

Basisphenoid of *Ophthalmosaurus* and *Ichthyosaurus*: A and B, basisphenoids of *Ophthalmosaurus* from below; C, basisphenoid of *Ichthyosaurus* from below. (A, R. 2164; B, R. 2161; C, R. 2063: ⅔ nat. size.)

i.c.f., internal carotid foramina; *pas.*, parasphenoid; *pt.f.*, facets for pterygoids; *pt.p.*, pterygoid processes; *v.c.p.*, lower cylindrical processes.

and by the anterior angles of the pterygoid processes (*pt.p.*). As already mentioned, the middle of the ventral face is pierced by the internal carotid foramen (*i.c.f.*). In front of this the adherent posterior end of the parasphenoid is seen. In different individuals the degree to which the parasphenoid overlaps varies (see text-fig. 6).

OPHTHALMOSAURUS.

Text-fig. 7.

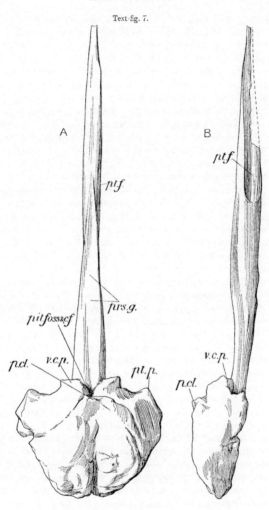

downwards: A, from above; B, from side. (R. 2180, ⅔ nat. size.)

p.cl., posterior clinoid processes; *pit.foss. & c.f.*, pituitary fossa and carotid foramen; *prs.g.*, presphenoidal groove; *pt.f.*, facet for pterygoids; *pt.p.*, pterygoid processes of basisphenoid; *v.c.p.*, lower cylindrical processes of basisphenoid.

In one case (fig. 6, A) it extends back round the sides of the carotid foramen (*i.c.f.*), forming the anterior and part of the lateral border of that opening; in most (text-fig. 5, A) it extends just to the anterior border of the foramen, while in one instance it only extends about halfway to it (text-fig. 6, B).

In front of the basisphenoid the *parasphenoid* (text-fig. 7) extends forwards as a long pointed rostrum, the length of which is about three times that of the basisphenoid in the mid-ventral line. Posteriorly this rostrum is transversely oval in section, but soon becomes triangular, the upper surface being concave from side to side, so that it forms a shallow groove (*prs.g.*) which received the lower edge of the presphenoidal bar of cartilage, which probably, as in *Hatteria*, formed the ventral edge of the interorbital septum : this groove occupies about the posterior half of the free portion of the bone. In front of this the rostrum becomes more compressed laterally, and on its sides are long surfaces (*pt.f.*) slightly concave and ridged, apparently for union with the inner side of the anterior limbs of the pterygoids, between which its anterior half extended.

The *squamosal* (text-fig. 8, A, B.) is a bone of complex shape. It occupies the postero-superior angle of the skull, where it forms a prominent boss (*p.e.a.*). From this angle there runs forwards a broad plate, the somewhat thickened upper border of which forms the posterior half of the outer border (*o.b.*) of the supratemporal fossa (*s.t.foss.*); the ventral and anterior edges of this plate are thin and no doubt united with the supratemporal and postfrontal bones, as they are shown to do in Gilmore's figure * of the skull of *Baptanodon*, and as is the case in the earlier Ichthyosaurs. From the posterior angle again there runs inwards and a little forwards a very stout bar of bone, making an acute angle with that just described and forming the outer part of the posterior border of the supratemporal fossa. At its inner end this process widens out considerably from above downwards and terminates abruptly in a deeply hollowed, somewhat diamond-shaped cavity (*pa.f.*) for the reception of the outer end of the squamosal process of the parietal. On the posterior face of this process of the squamosal there is a small shelf-like projection, making nearly a right angle with the quadrate region about to be described: in this angle is the facet for the reception of the outer end of the opisthotic (*op.f.*). Beneath the posterior angle (*p.e.a.*) there is a broad plate of bone continuous in front and behind with the processes already described: this is the quadrate region of the squamosal (*q.f.*); it consists of an outer and an inner plate separated by a deep narrow fossa, into which the upper end of the quadrate is firmly fixed. The inner plate extends down the inner face of the quadrate, to which it is closely adherent, and at its lower end it overlaps the ascending quadrate plate of the pterygoid ; by this arrangement the quadrate, apart from its other supports, the stapes and quadrato-jugal, is held rigidly by the squamosal above and the pterygoid below. The above description agrees in the main with that given by Gilmore in his account of the skull of *Baptanodon*, but it cannot be said

* "Osteology of *Baptanodon* (Marsh)," Mem. Carnegie Museum, vol. ii. (1905) pl. viii.

Text-fig. 8.

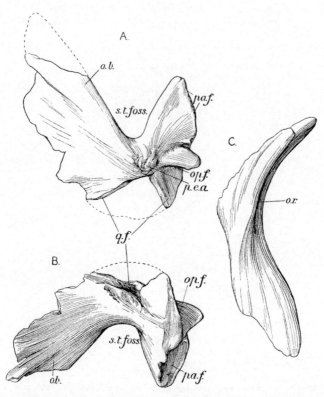

Squamosal and postorbital of *Ophthalmosaurus*: A, left squamosal from above; B, ditto from below (the dotted line represents the restored outline of the quadrate border); C, right postorbital. (A, B, R. 2146; C, R. 2180 : ⅔ nat. size.)

o.b., outer branch of squamosal; *op.f.*, facet for opisthotic; *o.r.*, orbital rim on postorbital; *pa.f.*, surface for union with the parietal; *p.e.a.*, postero-external angle of squamosal; *q.f.*, surface for union with upper end of quadrate; *s.t.foss.*, supratemporal fossa.

that the inferior plate filled all the space between the opisthotic, stapes, and quadrate, nor does its lower end seem to extend so far down as to pass under the upper lateral margin of the stapes. It seems probable that in the skull figured by Gilmore the squamosal and quadrate have been somewhat dislocated from their natural positions with regard to one another.

The *supratemporal* is missing or crushed beyond recognition in most cases, but in

Text-fig. 9.

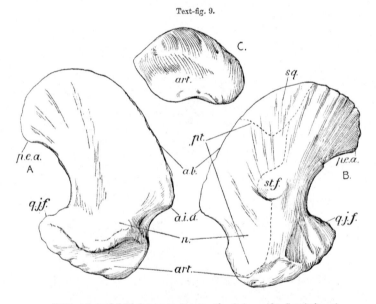

Right quadrate of *Ophthalmosaurus*: A, outer side; B, inner side; C, articular end.
(R. 2133: ⅔ nat. size.)

a.b., anterior border; *a.i.a.*, antero-internal angle; *art.*, articular surface for mandible; *n.*, neck of bone; *p.e.a.*, postero-external angle; *pt.* surface of union with pterygoid; *q.j.f.*, facet for articulation with quadrato-jugal; *sq.*, surface for union with the squamosal; *st.f.*, facet for outer end of stapes. (The outlines of the surfaces for the squamosal and pterygoid are shown by dotted lines.)

one (R. 2740) it can be seen that it was roughly triangular, overlapping the postorbital anteriorly, and sending forwards along it a process to meet the postfrontal. The relations with the bones behind and above cannot be made out.

The *quadrate* (Pl. I. fig. 10; text-fig. 9) is a large, broadly sickle- or rather ear-shaped bone, consisting of a comparatively thin upper portion and a greatly thickened articular region. The form of the articulation (text-fig. 9, C) varies somewhat, according to the degree to which the ossification of the cartilage has proceeded; in the most fully ossified specimens the surface is strongly convex from before back. In the transverse direction there is a double curve, the inner portion being convex, especially towards the middle of its length, while the outer is concave, especially in front. The prominent postero-external angle of this region forms a well-marked projection ($q.j.f.$), with the upper end of which the quadrato-jugal articulates by a concave surface (qf, text-fig. 10, A–C). A little above the articular end the bone narrows to form a sort of wide neck ($n.$): the anterior border of this region, when fully ossified, is sharp and gently concave; the posterior border is thickened and rounded and forms the lower part of the deeply concave posterior border, the degree of concavity varying greatly in different individuals. The upper and anterior border of the blade ($a.b.$) forms a long convex curve forwards to the upper end of the neck, with which it makes an obtuse angle ($a.i.a.$): the edge is thickened along the upper end and thinner anteriorly; throughout its length it seems to have been bordered with cartilage. The outer face of the blade is concave in all directions, and its posterior upper angle ($p.e.a.$) is bent sharply outwards; the inner face is divided into a postero-superior portion bent outwards, and a nearly flat anterior region which is overlapped by the squamosal above and the pterygoid below. The pterygoid is intimately united to the roughened surface marked in the figure by a dotted line ($pt.$); it terminates above by a thin tongue-like expansion, which in its turn is overlapped by the squamosal ($sq.$). Immediately behind the pterygoid plate, and in fact forming a notch on its posterior border, is the facet ($st.f.$) with which the outer end of the stapes unites. The squamosal, as already mentioned, embraces the upper end of the bone and on the inner face overlaps it and part of the pterygoid. The whole arrangement of the quadrate (see text-fig. 4) seems to be directed towards the attainment of the greatest possible rigidity: thus, on the inner side it is supported by the pterygoid, which, in its turn, is closely united with the basisphenoid; by the squamosal, which is joined to the parietal, supratemporal, and postfrontal, and is further braced by the outer end of the opisthotic and by its overlap on to the pterygoid; by the stapes, which extends from the above-mentioned notch to the basioccipital, and is further held in position by the inferior plate of the pterygoid, on which it rests. On the outer side also the quadrate is supported by the quadrato-jugal, though in this case the mode of union seems to have allowed a little movement between the two bones. The arrangement in the Rhynchocephalia is, of course, similar in many respects, but neither in that nor in any other group of reptiles does it appear that the stapes has been converted into a mere support for the quadrate.

The *quadrato-jugal* (text-fig. 10, A–C) is a small triangular bone, the postero-inferior angle being obtuse: both the posterior and lower borders are thickened, while the anterior (upper) edge is thin and sharp. The posterior lower angle is occupied by a rounded articular knob, the lower surface of which bears a rather deep cup-like depression ($q.f.$), into which the upwardly projecting postero-external angle of the quadrate is received. As already mentioned, there seems to have been a certain amount of play between the two bones; and this probably indicates that the rostral portion of the skull was capable of some small degree of movement on the cranial, this movement being transmitted to the quadrato-jugal by the jugal bar. The external face of the bone is overlapped below by the posterior upturned end of the jugal, the surface for which is clearly defined (see text-fig. 10, A, $ju.f.$). Above, the bone bears a roughened facet ($p.orb.f.$) by which it joined the postorbital, which also overlapped the upper end of the jugal.

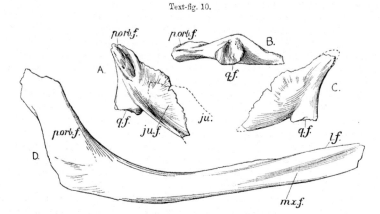

Text-fig. 10.

Right quadrato-jugal and jugal of *Ophthalmosaurus*: A, quadrato-jugal, outer side; B, ditto, lower side, with articulation for quadrate; C, ditto, inner side; D, jugal, outer side. (R. 2180, about ¾ nat. size.)

ju., jugal; *ju.f.*, facet for jugal; *l.f.*, facet for posterior end of lachrymal; *mx.f.*, facet for maxilla; *p.orb.f.*, facet for postorbital; *q.f.*, facet for quadrate.

The *jugal* (text-fig. 10, D) is a long curved bone, consisting of a slender horizontal bar in front and a laterally-flattened, upturned blade behind. The upper portion of the blade overlaps the outer face of the quadrato-jugal, and is in turn overlapped by the lower end of the postorbital ($p.orb.f.$), so that it is thrust between these two bones. The anterior bar is oval in section in its posterior half, but anteriorly it becomes compressed from side to side, and is deeply grooved ($l.f.$) superiorly for union with the lower edge of the posterior prolongation of the lachrymal. On the inner side of this region there is an elongated and well-defined surface, which is flat or gently concave, for union with the posterior limb of the maxilla. The jugal formed the whole of the ventral and the lower part of the posterior border of the orbit.

The upper part of the posterior border of the orbit is formed by the long curved *postorbital* (text-fig. 8, C), the lower end of which widens out a little and overlaps the outer face of the upper portion of the jugal. Above this the bone bears on its inner face a surface for union with the upper end of the quadrato-jugal, and above this again its upper face is overlapped first by the supratemporal, then by the postfrontal. In this upper region the outer surface bears a prominent ridge, which forms the actual rim of the orbit.

Text-fig. 11.

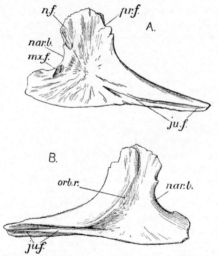

Right lachrymal of *Ophthalmosaurus*: A, inner side; B, outer side. (R. 2180, ⅔ nat. size.)
ju.f., facet for jugal; *mx.f.*, facet for maxilla; *nar.b.*, border of external nares; *n.f.*, facet for nasal; *orb.r.*, rim of orbit; *pr.f.*, facet for prefrontal.

The *lachrymal* (text-fig. 11) is an irregularly triangular bone, with the posterior angle produced backwards into a process, of which the ventral surface is raised into a strong ridge internally and a deep groove externally (*ju.f.*), both fitting on to corresponding surfaces on the upper side of the anterior end of the jugal. The upper border of this process is continued backwards and upwards on to the outer face of the bone as a curved ridge (*orb.r.*) running towards the upper angle and forming a rim to this part of the orbit. The upper angle is blunt and rounded, and bears on its antero-internal face a surface (*n.f.*) for union with the nasal. A roughening (*pr.f.*) on the posterior edge probably marks the junction with the prefrontal, but this is not clear in the specimens described. The anterior concave border (*nar.b.*) of the bone is rounded and reflected inwards, forming the posterior border of the external nares. On the inner face of the bone, at the lower end of the nasal border, is a rugose surface (*mx.f.*) for union with the maxilla, to the outer face of which the lower part of the lachrymal is closely applied.

The *nasals* (text-fig. 12) are very large bones of peculiar form. They are prolonged forwards to a great extent, their anterior portion being concealed beneath the premaxillæ, to the inner face of which they are here closely applied: they extend forwards to a point about two-thirds of the total length of the rostrum in front of the anterior border of the nostrils. Further back, by the divergence of the premaxillæ, they are

Text-fig. 12.

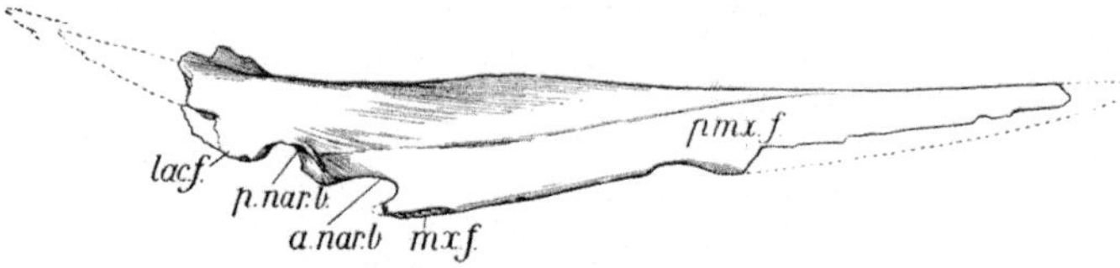

Right nasal of *Ophthalmosaurus* from outer side. (R. 2180, about ⅔ nat. size.)

a.nar.b., anterior border of external nares; *lac.f.*, facet for lachrymal; *mx.f.*, facet for maxilla; *pmx.f.*, facet for premaxilla; *p.nar.b.*, posterior portion of border of external nares.

exposed on the surface of the snout, and extend backwards between the nasal openings to behind the level of the front of the orbit. In this posterior region each is divided by a rounded angle into two regions: (1) an upper, which with its fellow of the opposite side forms the roof of the snout, which is somewhat concave between and behind the narial openings; (2) a lateral portion forming the upper border of the nasal opening, which is divided into an anterior portion where this upper border (*a.nar.b.*) is thin and sharp, and a posterior region (*p.nar.b.*), separated by a process projecting into the opening, where the edge is inflected to form a sort of funnel-shaped channel. Immediately in front of the nasal opening the lower border of the bone bears an elongated roughened facet for union with the corresponding surface of the maxilla, which bounds the opening below. Just behind its narial border the nasal has a roughened surface (*lac.f.*) for junction with the lachrymal, by which the posterior border of the opening is formed. No specimen in which the posterior end of the nasal is well preserved is known, but it seems to have been in contact with the frontal and prefrontal bones. The inner face of the bone, which is concave from above downwards, is marked by a series of longitudinal ridges running forwards from the nasal opening.

The *maxilla* (text-fig. 13) is a small bone compared with the premaxilla and in no case seems to have borne teeth, though the dental groove begins in its anterior portion. The bone, as a whole, is a shuttle-shaped structure, with the upper and inner surfaces deeply concave from above downwards. The posterior prolongation bears on its outer face (text-fig. 13, A) an elongated concave, grooved surface (*ju.f.*) for

Text-fig. 13.

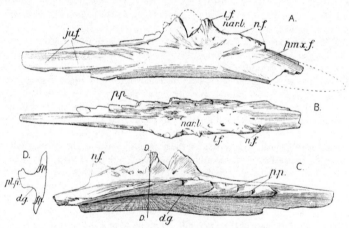

Right maxilla of *Ophthalmosaurus*: A, outer side; B, upper side; C, inner side; D, vertical section at the line D–D in fig. C. (R. 2180, ½ nat. size.)

d.g., dental groove; *f.p.*, facial portion of bone; *ju.f.*, facet for jugal; *l.f.*, facet for lachrymal: *nar.b.*, border of external nares; *n.f.*, facet for nasal; *pl.p.*, palatine plate; *pmx.f.*, facet for premaxilla: *p.p.*, processes on palatine border.

union with the overlapping jugal: there may have been a little play between the two elements. In front of this the outer face of the bone is flat above and gently convex below: its upper border is produced upwards into a blunt triangular process, the outer face of which is overlapped by the lachrymal, with which it unites in a short strong suture (*l.f.*) along its upper edge. In front of this projection the upper border of the bone is rounded and forms part of the lower edge of the nasal opening (*nar.b.*). Anterior to this, again, there is an elongated and roughened facet (*n.f.*) for union with the corresponding surface on the lower edge of the nasal. The anterior end of the bone runs out into a pointed process, the outer face of which is overlapped by the pointed posterior prolongation of the premaxilla. Looked at from above (text-fig. 13, B), the bone is seen to widen out considerably towards its middle, forming a palatine plate which is concave from side to side: possibly this concavity, anteriorly at least, received a posterior process of the premaxilla. The upper edge of the inner border of the palatine plate is peculiarly irregular, being produced into a series of elongated processes which lie almost parallel with the long axis of the bone; possibly these were connected with the bones of the palate, but their use is obscure. Beneath the palatine expansion ($pl.p.$) the inner face of the bone is deeply concave from above downwards, and anteriorly forms the posterior portion of the dental groove ($d.g.$), in which, however, no trace of the presence of teeth can be observed.

The *premaxillæ* are greatly enlarged and elongated bones, forming the greater part of the rostrum. Anteriorly they terminate in a blunt point, from which they gradually increase in depth backwards: their outer surface is convex from above downwards, and a little above the alveolar border bears a strongly-marked longitudinal groove, into which numerous foramina, probably vascular, open; this groove begins a little in front of the narial opening and dies away a few centimetres from the end of the snout, though the foramina are continued right to the end. The palatal face of the bone bears the broad alveolar groove continuous posteriorly with that of the maxilla. On the outer side of the anterior portion of the groove is a series of regular indentations separated by rounded ridges and marking the points of insertion of the relatively small and apparently very loosely attached teeth. The inner side of the groove is formed by a thickened rounded border, which, in the anterior region at least, was in contact with its fellow of the opposite side, but posteriorly the two were probably separated by the vomers, though the structure of this part of the skull is not well shown in any specimen. The inner face of the premaxilla is concave from above downwards, forming a long channel-like groove deepening from before backwards. For some distance this groove receives the anterior end of the nasals, as noticed above, and for a long distance the premaxillæ meet above those bones. For the greater part of their junction in the mid-dorsal line the premaxillæ seem to have been merely in contact, but just before they diverge they were united by a strong ligamentous connection, the remains of which are seen in the elongated roughened surface occurring at this point. Behind this the superior border slopes away towards the inferior, the two meeting at the end of the long pointed posterior prolongation which overlaps the outer face of the maxilla. This posterior prolongation may have been slightly notched by the anterior angle of the narial opening, as in *Ichthyosaurus acutirostris* (see Fraas, 'Die Ichthyosaurier,' pl. ii. fig. 1).

The *parietal* (text-fig. 14) is a bone of very peculiar shape. It consists of a body which formed the posterior part of the roof and part of the side-wall of the cranium,

OPHTHALMOSAURUS.

Text-fig. 14.

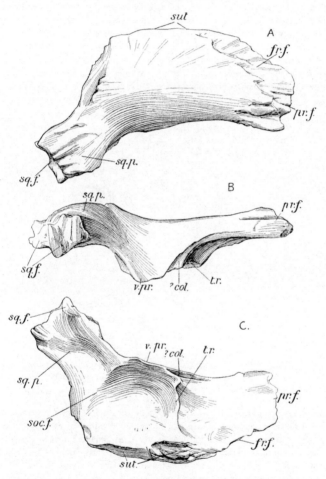

Right parietal of *Ophthalmosaurus*: A, from above; B, from outer side; C, from below. (R. 2162, ⅔ nat. size.)

? col., depression which probably received the upper end of the *columella cranii*; *fr.f.*, facet for frontal; *pr.f.*, facet for postfrontal; *soc.f.*, surface for supraoccipital; *sq.f.*, facet for squamosal; *sq.p.*, squamosal process; *sut.*, suture between parietals; *t.r.*, tentorial ridge across roof of brain-case; *v.pr.*, ventral (outer) process.

and of a process running outwards and backwards to meet the squamosal, with which it unites in a strong complex suture; the upper surface of this process is convex from before backwards, the ventral concave. The upper surface of the body of the bone is strongly convex from side to side; posteriorly the outer border is produced downwards to the side of the brain-case, while anteriorly it bears on its lower edge a deep groove into which probably the upper end of the epipterygoid fitted. The inner side bears a flat sutural surface for union with the corresponding bone of the opposite side: this sutural surface consists of a comparatively narrow posterior portion and a broad

Text-fig. 15.

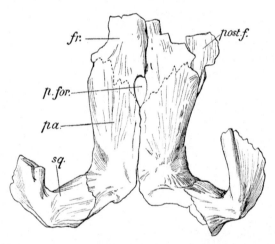

Part of upper region of skull of *Ophthalmosaurus*, showing the squamoso-parietal bars and the skull-roof with the parietal foramen. (R. 3535, ½ nat. size.)

fr., frontal; *pa.*, parietal; *p.for.*, pineal foramen; *post.f.*, postfrontal; *sq.*, squamosal.

oval roughened anterior region, which is borne on the inner end of a prominent ridge which crosses the cranial surface somewhat obliquely, dividing it into distinct anterior and posterior halves. The posterior half is about as broad as long, concave in all directions, while the anterior is narrower and gently concave from before back only. It appears that this ridge formed a sort of tentorium across the roof of the brain-case, and probably marks the line of separation between the optic lobes and the cerebral hemispheres, the former probably being very large. In front of the median suture the inner border of the bone slopes away from the middle line and it is overlapped to a considerable extent by the frontals; it appears that the parietal may or may not take part in the formation of the pineal foramen. External to the surface for the frontals there is another deeply ridged suture, probably for the prefrontal. The line of union between the parietal and the supraoccipital is a not very well-marked ridge, and the junction of the two bones seems to have been weak, a pad of cartilage probably intervening between them. The rest of the bones of the skull-roof are badly preserved in all the

Text-fig. 16.

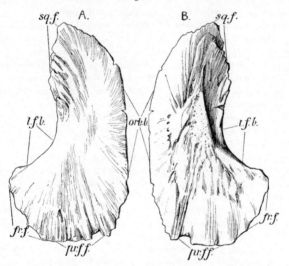

Left postfrontal of *Ophthalmosaurus*: A, from above; B, from below. (R. 2146, ⅔ nat. size. *fr.f.*, surface of union with the frontal and (?) parietal; *orb.b.*, orbital border; *pr.ff.*, surface of union with the prefrontal; *sq.f.*, ditto for squamosal; *t.f.b.*, border of temporal fossa.

available specimens, so that they cannot be described in detail. The *frontals* (text-fig. 15) are small oblong bones, the postero-internal angle being in most cases cut away to form the border of the pineal foramen, which is a large oval aperture; in front of this they unite in the middle line by a strong suture. Externally they unite behind with the postfrontals, and in front apparently with the prefrontals and nasals. The *postfrontal* (text-fig. 16) is large; it widens out towards each end. Its relations posteriorly cannot be made out clearly, but it seems to have united behind in a complex suture with the squamosal; anteriorly it joins the frontal, prefrontal, and perhaps the parietal. In its narrow middle region it forms the upper border of the orbit and the outer border of the supratemporal fossa; its orbital border is thin and sharp, and within this the bone thickens gradually towards its middle to form a gently concave roof to the orbit. The inner edge forming the border of the supratemporal fossa is thickened and rounded.

The *prefrontal* consists of a strong rounded central portion, which is expanded above to form the rim of the orbit. At the lower decurved end it interlocks with the upper end of the lachrymal. On its upper surface it bears a strong ridge, which fits into a corresponding groove on the overlapping nasal. Postero-internally it joins the frontal, and externally its upper surface is extensively overlapped by the postfrontal.

The *pterygoids* (text-fig. 17) are very large bones, expanded posteriorly, and produced anteriorly into long pointed processes. The broad posterior portion united by its inner upper surface (*bs.pt.f.*) with the basipterygoid processes of the basisphenoid, and behind these also overlapped the ventral surface of that bone to a considerable extent by a postero-internal expansion (*in.p.*). External to this expansion and separated from it by a deep groove the dorsal surface of the bone bears a strong oblique crest-like ridge (*d.p.*), the upper surface of which, together with that of the postero-external process (*e.p.*), is closely adherent to the inner face of the quadrate, the posterior half of which it completely covers; at its upper end it is itself overlapped for a short distance by the quadrate process of the squamosal. Into the deep groove (*g.st.*) between the dorsal and postero-external processes the lower border of the large stapes lies, its outer end, as already described, fitting against a facet of the inner face of the quadrate immediately behind the pterygoid. In front of the posterior expansion the bone first narrows into a sort of neck, of which both the outer and inner borders are concave; then it expands into a broad flat plate, which gradually narrows forwards till, at about one-third of its length from the anterior extremity, it passes forwards into an elongated vertical plate forming the anterior third of the bone. This plate bears on both its outer and inner surfaces elongated slightly ridged facets for union with adjoining bones, that on the outer face being no doubt for the vomer, that on the inner for the corresponding facet on the pterygoid of the opposite side. The posterior third of the inner border (*i.b.*) of the portion of the bone in front of the union with the basisphenoid is thickened and concave; it forms the outer border of the interpterygoid vacuity, which is divided in the middle line by the parasphenoid, the anterior end of which thrusts itself between and unites with the pterygoids for some distance (*pas.f.*); in this region the inner border of the pterygoids is nearly straight. Anteriorly, as already mentioned, they are in contact with each other in the middle line, and probably lay between the vomers. The thickened inner border of the posterior part of the bone forms a ridge (*r.*), which is continuous anteriorly with the upper edge of the vertical intervomerine plate.

OPHTHALMOSAURUS.

The *palatines* are not well known. The bone figured in text-fig. 18 is believed to be one of these elements, since in its general characters it is similar, so far as can be seen, to the palatine of *Ichthyosaurus zetlandicus*. The figure is drawn mainly from

Text-fig. 17.

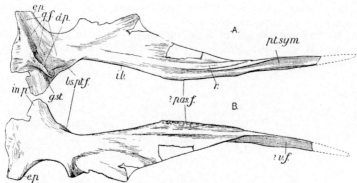

Left pterygoid of *Ophthalmosaurus*: A, upper surface; B, lower (palatal) surface. (R. 2150, ¼ nat. size.)

bs.pt.f., facet for union with the pterygoid process of the basisphenoid; *d.p.*, dorsal process; *e.p.*, outer process; *g.st.*, groove for the reception of the stapes; *i.b.*, internal border; *in.p.*, inner process; *? pas.f.*, facet for union with the parasphenoid (?); *pt.sym.*, symphysial surface for opposite pterygoid; *q.f.*, surface for union with the quadrate; *r.*, dorsal ridge; *v.f.*, facet for vomer (?).

Text-fig. 18.

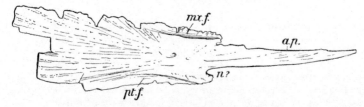

? Right palatine of *Ophthalmosaurus*, from below. (⅔ nat. size.)

a.p., anterior process; *mx.f.*, facet for maxilla; *pt.f.*, facet for pterygoid; *n?*, border of internal nares.

the right bone, but the posterior portion has been restored from that of the left side. If this interpretation be correct, the bone would seem to consist of a long anterior process (*a.p.*), the inner rounded border of which formed the outer side of the internal nares, while its thinner outer edge united with the premaxillæ. Posteriorly the bone is broader, and on both its outer and inner borders has shelf-like facets for union with the neighbouring bones, the outer probably joining the maxilla and premaxilla (*mx.f.*), the inner meeting the vomer in front and the pterygoid behind. The notch marked *n.?* in the figure is apparently the posterior angle of the internal nares. It is only fair to state that in one specimen, including many parts of the skull, there is a bone probably belonging to the palate, which is of considerably different form, and may be a palatine. In fact, till a skull is found with the palate preserved in something like its original condition, it will remain impossible to be certain of the form and relations of its constituent elements. This remark applies especially to the vomers. The bones that

Text-fig. 19.

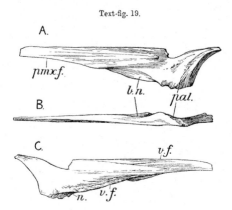

? Left vomer of *Ophthalmosaurus*: A, outer side; B, palatal surface; C, inner side. (R. 3533, ⅔ nat. size.)
b.n., border of nasal opening; *n.*, narial groove; *pal.*, surface of union with the palatine;
pmx.f., surface for union with the premaxilla; *v.f.*, surfaces of union with the opposite vomer.

are here regarded as *vomers* are figured in text-fig. 19; they belong to a small skull, of which most of the bones are preserved in a separate condition. The form in side view is shown in text-fig. 19, A and C, while the narrow palatal surface is seen in 19 B. The outer face bears on its anterior pointed region a facet probably for union with the premaxillæ, while on its posterior and lower border is a surface which may have united with the palatine (*pal.*). The rounded notch marks the inner border of the internal nares. The outer face in front of the narial notch (*n.*) bears along its upper and lower edges facets for union with the vomer of the opposite side (*v.f.*). On the whole, these bones agree very well with what can be seen of the vomers in some Ichthyosaurian skulls, and may be regarded with some confidence as being those elements.

Sclerotic Ring (Pl. I. figs. 9, 10).—The large and strongly ossified sclerotic ring in the Ichthyosaurian eye is one of the most notable characteristics of the group, but nevertheless does not seem to have been satisfactorily described. In the present collection there are several more or less nearly complete specimens of this structure, some with a greater or less number of sclerotic plates still united, others in which they have all become separated. From these it appears that the plates not only formed a ring round the pupil on the front of the eye, but externally also curved rather sharply round on to the back of the eyeball, over which they extended some distance, though not so far as in front. Owen[*] has already described this in the case of some Liassic Ichthyosaurs, and points out that the structure of the ring indicates "the extreme oblateness of that visual spheroid." Each separate plate extends from front to back, of course narrowing towards the pupil; in the young the plates are very thin, but they become considerably thickened with advancing age. They unite along their edges in a complex suture (Pl. I. fig. 10), interlocking in such a way that no movement can have taken place between them. Certainly in no specimen have they been seen to overlap simply one another as has been figured by Gilmore in the case of *Baptanodon*[†].

Even in birds, especially the Hawks and Owls, in which the plates often imbricate, they frequently fit into grooves in one another's edge, and in any case are so firmly united by connective tissue that little or no movement between the separate elements is possible. As already noticed, the plates are very thin in the young; they seem to ossify from a centre lying just in front of the point where they bend round to the back of the eye, and at this centre of growth there is a rugose surface from which there radiates in all directions a series of fine lines, which give the bone a fibrous appearance. Growth in thickness takes place by the addition of successive laminæ, probably on the inner side only; the edges of these laminæ of two contiguous plates interlock in the manner above described.

Mandible (figs. 20-23).—The structure of the mandible, on the whole, agrees very closely with that described by Gilmore in the case of the American forms. Each ramus consists of six elements—one, the articular, being a cartilage-bone, the remainder sheathing membrane-bones. Of these latter the *dentary* (*dent.*) is the largest: anteriorly it forms the whole of the jaw and terminates in a point; posteriorly it runs back, overlapping the anterior ends of angular and surangular on the outer face of the ramus, and terminating in a point a little in front of the coronoid process of the surangular referred to below; internally it extends back about the same distance as externally, but in the specimens examined its posterior end is concealed by the overlapping coronoid bone. The outer face is convex from above downwards, and a little

---

[*] 'Rept. Lias. Form.' (Mon. Pal. Soc. 1881) pt. iii. p. 103.
[†] "Notes on Osteology of *Baptanodon*," Mem. Carnegie Museum, vol. ii. (1906) p. 328, fig. 3.

below and parallel with the alveolar border is marked by a deep groove, into which several foramina open; anteriorly this groove is replaced by a number of irregularly placed foramina. On the inner side the dentaries of opposite sides unite in the middle line for a little more than the anterior half of the total length of the symphysis, being separated in the posterior portion by the splenials. In their symphysial region the dentaries unite by their flattened upper and lower borders, between which they are concave from above downwards. The surface covered by the splenial seems to have been nearly flat, and posteriorly, as already mentioned, the thin posterior extension of the bone is wedged in between the coronoid and surangular. The upper surface of the dentary bears the broad and not very deep alveolar groove, in which the presence

Text-fig. 20.

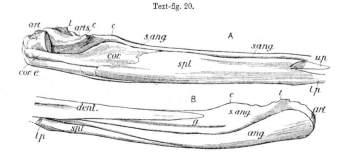

Posterior portion of left ramus of mandible of *Ophthalmosaurus*: A, inner side; B, outer side. (R. 2180, ⅓ nat. size.)

ang., angular bone; *art.*, articular bone; *art.s.*, articular surface on surangular; *c.*, *c'.*, coronoid-like processes on surangular; *cor.*, coronoid bone; *cor.e.*, backward extension of coronoid bone beneath the articular; *dent.*, dentary bone; *g.*, groove on outer face of surangular; *l.p.*, lower process at anterior end of the splenial bone; *s.ang.*, surangular bone; *spl.*, splenial bone; *t.*, tuberosity on upper border of surangular; *u.p.*, upper process at anterior end of the splenial bone.

of teeth is indicated by a number of transverse ridges, strongly marked in front but becoming less distinct further back. The outer border of the groove is comparatively broad and rounded, the inner thinner and sharper—particularly posteriorly, where it is reinforced by the upper border of the splenial, and by which still further back it is replaced.

The *splenial* (*spl.*) is a very large bone, which, as above mentioned, extends into the posterior half of the symphysis: at its thin anterior end it is divided by a large cleft into an upper and lower process, the inner face of each bearing a roughened surface for union with its fellow of the opposite side; probably the posterior end of the cleft represents the foramen figured by Gilmore[*], but if so it was entirely concealed within the symphysis. The upper border of the bone is rounded and helps to form the inner wall of the posterior part of the alveolar groove. Behind this it thins, and its upper border slopes away till it terminates in a point about the level of the coronoid process. The outer face is applied to the dentary in front, and to the coronoid and angular behind. Ventrally the anterior portion of the bone appears below the dentary and forms the lower edge of the mandible, but farther back it thins out and its lower border slopes up to its posterior termination, the posterior third or so of the bone lying entirely on the inner face of the ramus.

The *surangular* (*s.ang.*, text-figs. 20, 21) forms the upper border of the hinder portion of the mandible. It is a long bone, consisting of a pointed anterior prolonga-

Text-fig. 21.

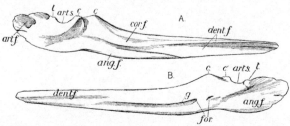

Left surangular bone of *Ophthalmosaurus*: A, inner side; B, outer side. (R. 2740 about ½ nat. size.)

ang.f., facet for union with the angular; *art.f.*, facet for union with the articular bone; *art.s.*, articular surface; *c.*, *c'.*, coronoid processes; *cor.f.*, facet for union with the coronoid bone; *dent.f.*, surfaces of union with the dentary; *for.*, foramen on outer surface of bone; *g.*, groove running forwards from the foramen; *t.*, tubercle at hinder end of articular surface.

tion and a much broader posterior articular portion. It is partly concealed in front by the overlapping of the dentary, but below this the surface is exposed and bears a very deep longitudinal groove (*g.*) into which a foramen (*for.*) opens posteriorly. The inner face of the anterior portion is flattened and covered first by the dentary and behind this by the anterior expansion of the bone here called the coronoid (*cor.*); this, however, diverges from it posteriorly, and slopes away so that for some distance its inner face is exposed. Behind this point comes the articular bone (*art.*), closely adherent to the inner face of the surangular. The upper border of the bone is rounded in front, but posteriorly becomes thinner, and is raised into two blunt processes, one directed upwards, the other a little further back directed somewhat

---

[*] "Notes on Osteology of *Baptanodon*," Mem. Carnegie Museum, vol. ii. (1906) p. 327, fig. 1.

inwards; these seem to be functionally coronoid processes, the posterior one especially having roughened surfaces for muscle-attachments. Behind these again the upper border spreads out laterally and is deeply concave, the concavity being marked off posteriorly by a prominent tubercle (*t*), behind which is the thin upper edge of the expanded portion which, as above described, bears the articular on its inner face. Ventrally the surangular receives the upper border of the angular in front, while behind its outer face is extensively overlapped by that bone.

The *angular* (*ang.*) forms the lower border of the posterior part of the mandibular ramus; it consists of an elongated ventral portion, the upper edge of which, as just mentioned, interlocks with the ventral border of the surangular; below and internally it is closely united with the splenial. Posteriorly it widens out and forms the postero-inferior angle of the jaw; in this region it extensively overlaps the surangular

Text-fig. 22.

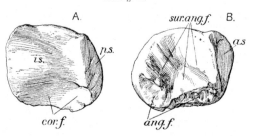

Right articular bone of *Ophthalmosaurus*: A, inner surface; B, outer surface. (R. 2180, ⅔ nat. size.) *ang.f.*, facet for the angular bone; *a.s.*, anterior surface; *cor.f.*, facet for the backward prolongation of the coronoid bone; *i.s.*, inner surface; *p.s.*, posterior surface; *sur.ang.f.*, facet for the surangular bone.

externally, while internally it is firmly united to two distinct facets on the lower edge of the articular (*q.v.*). On its inner side also its upper edge unites with the backward prolongation of the coronoid.

The *coronoid* (*cor.*) is situated entirely on the inner face of the ramus, and does not project above it, the prominences on the surangular apparently replacing the true coronoid process. In front the bone is deep and thin, and is wedged in between the dentary on the inner, and the surangular on the outer side. Farther back it narrows and is separated from the surangular by a deep fossa, which is closed behind by the anterior end of the articular. Posteriorly the bone sends back a long process to the hinder end of the jaw, wedged in between the lower side of the articular and the upper border of the angular.

The *articular* (*art.*, text-figs. 20, 22) is a cartilage-bone quite different in texture

OPHTHALMOSAURUS.

from the other elements of the mandible. Its inner free surface is convex from above downwards, and also from before backwards in its anterior half; posteriorly it is gently concave in the latter direction. Ventrally the inner surface is produced a little downwards, the inner concave surface (*cor.f.*) of the expansion receiving the posterior end of the coronoid. The gently concave anterior face (*a.s.*) is roughly semicircular in outline, the straight side being external and in contact with the surangular. The posterior end (*p.s.*) is slightly convex in all directions, and is narrowed below both internally and externally by the surfaces for the coronoid and angular respectively. Looked at from the outer side, the greater part of the face is occupied by a deeply concave facet for union with the surangular, while at the postero-

Text-fig. 23.

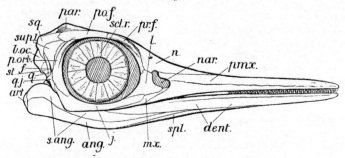

Restoration of the skull and mandible of *Ophthalmosaurus*. (About ⅓ nat. size.)

ang., angular; *art.*, articular; *b.oc.*, basioccipital; *dent.*, dentary; *f.*, foramen between the quadrate and quadrato-jugal; *j.*, jugal; *l.*, lachrymal; *mx.*, maxilla; *n.*, nasal; *nar.*, external nares; *par.*, parietal; *pmx.*, premaxilla; *po.f.*, postfrontal; *p.orb.*, postorbital; *pr.f.*, prefrontal; *q.*, quadrate; *q.j.*, quadrato-jugal; *s.ang.*, surangular; *scl.r.*, sclerotic ring; *spl.*, splenial; *sq.*, squamosal; *st.*, stapes; *sup.t.*, supratemporal.

inferior angle is another facet for union with the angular, though the most extensive junction with this bone is effected by a roughened surface truncating its inferior border. The articulation with the quadrate seems to have been formed by the anterior face of the articular and the outwardly-turned upper border of the posterior part of the surangular.

The above account agrees very closely with that given by Gilmore for *Baptanodon*, and still further confirms the probable identity of that genus with *Ophthalmosaurus*. Gilmore's figures[*] given in this second paper should be compared, especially his text-figures on p. 327 and plate xxxvi. of the work quoted below.

[*] "Notes on the Osteology of *Baptanodon*," Mem. Carnegie Museum, vol. ii. (1906) p. 325.

Dentition (Pl. I. figs. 1-8).—The dentition both in the upper and lower jaw seems to be undergoing reduction. This is shown by the relatively small size of the teeth, and, in the adult, by their apparent absence from the hinder part of the jaws, where no traces of alveolar divisions can be seen in the dental groove. In the young the front of the jaw bears numerous closely-set teeth, in one case as many as eleven in a space of 7 cm. (Pl. I. figs 7 and 8). In this case the teeth are in actual contact with one another, and are inclined a little backwards. In older jaws the teeth in nearly all cases have fallen out, their former position in the wide dental groove being marked by a series of alveolar pits, which are separated from one another by transverse rounded ridges. The latter are very distinctly marked on the front of the jaws, but become fainter and fainter as they are followed backwards, till they finally disappear altogether. Judging from the relative smallness of the teeth and the large size of the alveolar pits, it would appear that the teeth in the adult must have been very loosely implanted in a soft gum, the putrefaction of which after death would account for the detached condition in which the teeth are ordinarily found. In some cases, as in other Ichthyosauria, the roots of the teeth are marked by a deep pit resulting from the absorption of their substance before the advancing point of a replacing tooth.

The teeth themselves vary considerably in form, some being much curved, others nearly straight. The crowns (Pl. I. figs. 2 and 4-6) are sharply-pointed cones; they are usually curved and nearly circular in section; they are covered with a fairly thick coat of enamel, which in the middle portion of the crown is raised into longitudinal ridges which disappear towards the tip and the root. Beneath the enamel-clad crown the tooth thickens considerably, and the smooth dentine is exposed or at most covered with a very thin layer of cement (Pl. I. fig. 3); in this region there are sometimes rounded ridges running round the tooth. Lower down the dentine is thickly covered with cement, and is thrown into strong longitudinal folds; in some cases (Pl. I. fig. 1) these are so deep that their inner ends meet and enclose a small separate portion of the pulp-cavity. This cavity is very small in the crown, but lower down widens out and near the base is filled with a loose network of cement (see Pl. I. figs. 1 and 3).

The *hyoid arch* is represented only by a pair of elements (the ceratohyals ?). They are curved and flattened bars of bone, widening out gently towards their extremities, which are abruptly truncated at right angles by surfaces that appear to have been tipped with cartilage during life. Towards the ends of the bone the walls are frequently crushed in, as if the interior had been hollow or at least filled with comparatively soft tissue.

Vertebral Column (text-figs. 24-31).—In all the specimens hitherto examined the centra of the axis and atlas vertebræ are already fused together, though in some cases the line of junction is still plainly visible (text-fig. 24, D). The anterior face of the *atlas* is deeply cupped and is roughly triangular in outline, the angles being rounded off. The ventral angle is further truncated by a roughened surface (*w.b.f.*) looking

Text-fig. 24.

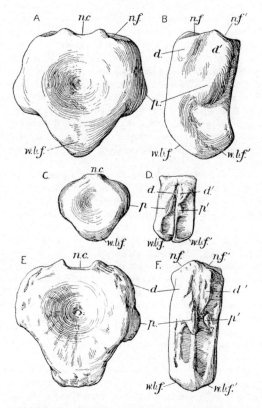

Centra of the atlas and axis of *Ophthalmosaurus*: A, anterior surface, and B, lateral surface of an uncrushed specimen (R. 2150 a); C, anterior surface, and D, lateral surface of a young specimen (R. 2175); E, anterior surface, and F, lateral surface of a somewhat compressed specimen (R. 2152). (All ? nat. size.)

$d., d'.$, diapophyses; $n.c.$, neural canal; $n.f., n.f'.$, surfaces for neural arches; $p., p'.$, parapophyses: $w.b.f., w.b.f'.$, facets for subvertebral wedge-bones.

downwards and forwards and marking the position of attachment of the anterior subvertebral wedge-bone. The upper border of the centrum is divided into three slight concavities, the median one forming the floor of the neural canal ($n.c.$), the anterior border of which is here usually rounded off, while the two lateral concavities ($n.f.$) are the surfaces for union with the neural arch. These latter surfaces are continuous with the diapophyses, the degree of development of which, as well as of the parapophyses, varies greatly in different individuals and even on opposite sides of the same atlas. Thus the diapophysis ($d.$, $d.'$) may be a clearly-defined, elongated, triangular surface, and the parapophysis ($p.$, $p.'$) an equally clearly-marked round tubercle, situated on the side of the centrum, nearer its posterior than its anterior edge : when this is so, the arrangement of the surfaces on the axis is similar (see text-fig. 24, F). In other specimens the parapophysis of the axis seem to have shifted forwards to the anterior edge of the centrum (text-fig. 24, D), and may unite with that of the atlas to form a single bony prominence. In some young examples the parapophysis may be confluent with the diapophysis.

As already mentioned, the *axis* is already fused with the atlas in the youngest specimens found, and no trace of the line of junction remains in the region forming the floor of the neural canal, though its position may be marked by a groove round the remainder of the centrum. The surfaces for the neural arch are separated from the corresponding facets of the atlas by transverse ridges, and are continuous externally with the diapophyses. As above noticed, the parapophysis may be confluent with that of the atlas, but more frequently is separable and situated near the posterior border of the centrum. The posterior face of this vertebra is roughly triangular and deeply concave : in some specimens there is a slight oblique truncation of the lower angle ($w.b.f.'$), which probably indicates that a subvertebral wedge-bone between this and the third cervical was present ; but neither it nor either of the anterior wedge-bones has been seen *in situ*, although that between the skull and the atlas was certainly present and that between the atlas and axis probably so. In the atlas and axis, as in the centra of the succeeding vertebræ, very great variability in length is observable in different individuals, and this is often so marked that it might easily be regarded as a character of specific value. Further examination shows that these differences are merely due to compression, the centra being frequently telescoped, as it were, till they may be little more than half their original length (see text-fig. 31, A, B), often at the same time showing little or no distortion in other directions. This seems to be the result of the giving way of the inner spongy tissue. Apart, however, from differences of form that can be thus explained, the variability in the shape of the vertebræ is very considerable ; but since intermediate forms between the extremes are usually to be found, it appears most probable that these differences may be regarded merely as individual variations, such as might easily occur in an imperfectly ossified skeleton in which much cartilage persisted throughout life.

OPHTHALMOSAURUS.

The centrum of the third cervical (text-fig. 25) is, like that of the axis, somewhat triangular, or perhaps rather pentangular in outline; it narrows somewhat ventrally, and bears on the mid-ventral line a fairly strongly-marked hypapophysial ridge. Both the anterior and the posterior faces are deeply cupped, and, in the type specimen at least, the whole of these faces, except a narrow rounded border, is concave; so that the statement that in this genus the concavity is confined to the central portion of the centrum and is surrounded by a flattened area, is not correct, except in the case of some specimens which have been somewhat crushed. The surfaces ($n.f.$) for union with the neural arch are deeply concave, and extend from end to end of the centrum; in front they are connected laterally with the diapophysial surfaces. The parapophyses are distinct tubercles, ending in a flattened or slightly concave surface; in this region

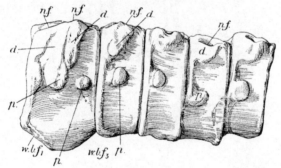

Centra of anterior cervical vertebræ of *Ophthalmosaurus*, from left side. (R. 2135, ⅔ nat. size.)
d., diapophyses; n.f., facets for neural arches; p., parapophyses; w.b.f., w.b.f.₃, facets for the first and third subvertebral wedge-bones.

they are situated not quite halfway down the side of the centrum and rather nearer the anterior than the posterior border. The next few vertebræ are very similar, except that at about the fifth their ventral surface becomes evenly rounded from side to side, the hypapophysial ridge disappearing; the degree to which this ridge is developed varies considerably in different individuals. As the vertebræ are followed backwards the centra become more circular in outline, the surfaces for the neural arch become longer and narrower, and the diapophyses separating from them (at about the 19th vertebra) shift lower and lower on the side of the centrum. This downward movement is also shared by the parapophyses, though to a less degree, so that about the 38th vertebra the diapophysis joins the papapophysis, the resulting faces being situated low down on the side of the centrum, the ventral face of which is here somewhat flattened. The vertebra in which the union of the two processes takes place is regarded as the first caudal. In the portion of the precaudal region in which the diapophyses are distinct

Text-fig. 26.

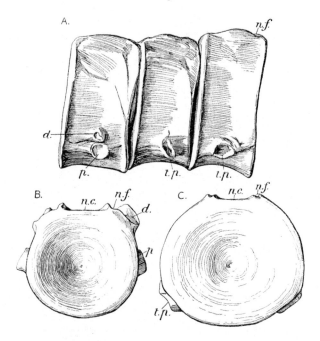

Dorsal and caudal vertebræ of *Ophthalmosaurus*: A, last dorsal and two anterior caudal vertebræ from the left side; B, anterior dorsal, posterior face; C, anterior caudal, posterior face. (Type specimen, R. 2133, ⅔ nat. size.)

d., diapophysis; *n.c.*, floor of neural canal; *n.f.*, facet for neural arch; *p*, parapophysis; *t.p.*, transverse process (combined diapophysial and parapophysial prominences).

from the neurapophyses, they are situated a little farther forwards than the parapophyses, and in the anterior part of this region they are nearer the anterior than the posterior border of the centrum. Throughout this region the anterior and posterior surfaces of the centra are fairly deeply concave, and although the sides of the concavity become steeper towards the centre, there cannot be said to be any flattened area round the periphery.

In the anterior caudal region (text-figs. 26, A, C, and 27, A, B) the form of the centra is very similar to that of the posterior dorsals, but, in addition to the general concavity of the anterior and posterior faces, there is often a deep pit in the middle of

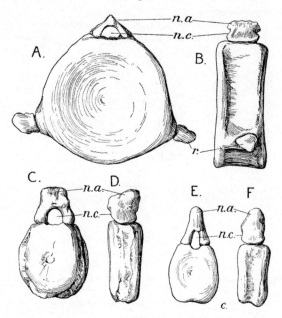

Caudal vertebræ of *Ophthalmosaurus*: A, anterior caudal from front; B, ditto from side; C, middle caudal vertebra (at bend of tail) from front; D, ditto from side; E, vertebra from just behind bend of tail, from front; F, ditto from side. (R. 3534, ⅔ nat. size.)

c., facets for chevrons; *n.a.*, neural arch; *n.c.*, neural canal; *r.*, rib.

the centrum, which is thus nearly perforated. In this region the ribs are small irregular nodules of bone (*r.*, text-fig. 27, A, B), and seem to have been of this form till they disappear altogether. In the middle caudals the single oblique rib-facet extends the whole width of the centrum, and appears to be higher up the side owing to the greater transverse convexity of the ventral surface. Above the rib-facet the sides are somewhat flattened and the neural surface is still nearly as broad as in the dorsal region; the rounded and as it were unfinished edges of the centrum show that a good deal of cartilage persisted in this region of the column, even in advanced life. Posteriorly the tail seems to have narrowed very rapidly, and the centra become first more rounded, then somewhat compressed laterally, so that they are deeper than wide, and at the same time the rib-facets disappear. This continues till two or three vertebræ of peculiar form (text-figs. 27, C, D, and 28, A, B) are reached. These seem to mark the point where the downward flexure of the column occurs; their centra are much deeper than wide and at the neural border are considerably longer than at the

Text-fig. 28.

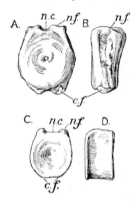

Caudal vertebræ of *Ophthalmosaurus*: A, centrum of vertebra at bend of tail from front; B, ditto from side; C, posterior caudal (some distance behind bend) from front; D, ditto from side. (R. 2169, ⅔ nat. size.)

c.f., facets for chevrons; *n.c.*, floor of neural canal; *n.f.*, facets for neural arch.

ventral. The concavities of the articular faces of the centra are surrounded by a broadly convex border; the middle of the depression bears a deep pit, marking the position of the notochord. The neural canal here forms a narrow deep groove, with the facets for the neural arch raised considerably above its floor. Both in these vertebræ and in a considerable number in front of them, the facets for the chevrons are well marked. Behind the bend of the tail come a large number of vertebræ in which ossification seems to be much more extensive (text-figs. 27, E, F, and 28, C, D) than in the anterior caudals, all the borders and surfaces being, as it were, neatly finished and sharply defined: this may result from the circumstance that these vertebræ were perhaps more mobile and more directly affected by definite strains and stresses from the outside. In them the centra are deeper than wide, evenly concave, without rounded borders, and have a neural surface wider than it is in some of the larger vertebræ in front; at the same time the neurapophysial ridges are less prominent. No pleurapophyses are present in this region.

In all cases the neural arch is found separated from the centrum clearly in consequence of the persistence of cartilage at the junction between the two; in a few instances it has been possible to refer the arches to the centra to which they actually belonged, but in most cases this is not possible.

In the first few cervical vertebræ the arch consists of two distinct halves, which

Text-fig. 29.

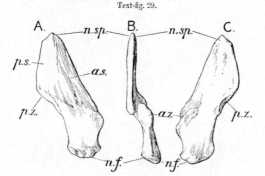

Left half of the neural arch of an anterior cervical vertebra of *Ophthalmosaurus*: A, inner side; B, from front; C, outer side. (R. 2180, ⅔ nat. size.)

a.s., anterior surface on inner face of neural spine for union with opposite half; *a.z.*, anterior zygapophysis; *n.f.*, facet for articulation with centrum; *n.sp.*, neural spine; *p.s.*, posterior surface of inner face of spine; *p.z.*, posterior zygapophysis.

unite above to form the neural spine, and it seems possible that there may have been a median element in this region, though this is not certain. Each of the two halves (text-fig. 29) consists of the pedicle terminating below in a rounded roughened facet for union with the centrum. This widens out above and bears the anterior and posterior zygapophyses, the former forming a strong prominence, while the latter forms an oblique surface on the postero-ventral border of the spine. This latter is a high compressed plate of bone, the outer face of which is smooth, while the inner bears two surfaces separated by a slight ridge: the anterior surface is roughened, apparently for union with the other half of the arch; the other surface indicates that this posterior portion overlaps on to the outer surface of the anterior half of the arch behind. The anterior border of the spine runs down as a strong crest on the inner side of the zygapophysis, which in this region was thus completely separated from its fellow of the opposite side. The posterior zygapophyses also seem to have been separated from one another, though not so clearly.

In the posterior cervical and dorsal regions the neural arches show no signs of division into two halves, and at the same time the anterior and posterior zygapophyses become confluent in the middle line, so that there is only a single anterior, and a single posterior median articular surface (text-fig. 30), as in the later species of *Ichthyosaurus*. The neural spine is a high plate of bone, sloping a little backwards and abruptly truncated almost at right angles at its upper end, which bears a shallow

Text-fig. 30.

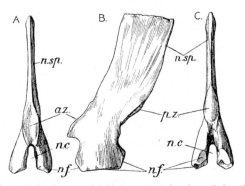

Neural arch of a dorsal vertebra of *Ophthalmosaurus*: A, from front; B, from side; C, from back. (R. 2137, ⅔ nat. size.)

a.z., anterior zygapophysis; *n.c.*, neural canal; *n.f.*, facets for union with centrum; *n.sp.*, neural spine; *p.z.*, posterior zygapophysis.

groove, apparently indicating that in life it was tipped with a crest of cartilage. The neural arch is high, but not very wide; the ends of the pedicles are deeply grooved, showing that they were capped with cartilage, the persistence of which, as above mentioned, fully accounts for the almost invariable separation of the arch from the centrum in the fossils.

In the middle caudal region the neural canal becomes very small and the arch very small, the spine being scarcely at all developed (text-fig. 27, A, B) and the zygapophyses entirely wanting. At the region where the downward bend takes place (text-fig. 27, C, D) the arches are stout ∩-shaped bones, the upper part of which forms a thick blunt spine projecting a little backwards. A little behind the bend (text-fig. 27, E, F) in the region of the tail-fin the arches become rather higher and more compressed, the spine sloping slightly backwards. The canal is small and is higher than it is wide; its lower portion is enclosed between the projecting processes of the centrum, with which the arch articulates.

The dorsal ribs (text-fig. 31, C, D) are long and comparatively stout. At the proximal end the capitular and tubercular facets are widely separated, the forking of the upper end being deeper than in most Liassic forms. The capitular facet (*h.*) is the larger and nearly circular in outline, while the smaller tubercular facet (*t.*) is

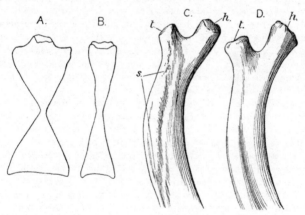

Sections of vertebral centra and the upper end of a dorsal rib of *Ophthalmosaurus*: A, vertical section through uncrushed vertebral centrum; B, vertical section through centrum of a crushed vertebra belonging to the same individual as last; C, posterior face of upper end of dorsal rib; D, anterior face of ditto. (A, B, R. 2180; C, D, R. 2137, ⅔ nat. size.)

h., head of rib; *t.*, tubercle of rib; *s.*, roughened ridge for muscle-attachment.

somewhat compressed from before backwards, its anterior border being nearly straight. The body of the rib is compressed from before back, but is so thickened along its outer side that both the anterior and posterior surfaces present the appearance of being grooved, the concavity being most marked on the posterior surface. On this surface near the upper end of the rib the upper border bears an elongated roughened surface (*s.*) for muscle-attachment. Towards their ventral ends the ribs become more rounded in section.

In the caudal region the single-headed ribs become reduced to mere nodules of bone (text-fig. 27, A, B), and finally disappear altogether a little in front of the point where the vertebral column bends downwards.

The ventral ribs seem to have been very slender and slightly developed, but they are not sufficiently known for description.

Shoulder-girdle and Fore Limb.—The shoulder-girdle of *Ophthalmosaurus* has already been described in some detail by Prof. Seeley [*], and, in fact, it is upon the characters of this part of the skeleton that the genus is founded. Unfortunately, however, the coracoids and one of the scapulæ of the type specimen are diseased and altogether abnormal, the left scapula, in fact, being an almost shapeless mass of bone, with which the upper end of the left clavicle is fused: probably this abnormality is due to extensive injuries received when the animal was comparatively small. The left coracoid, figured by Prof. Seeley, has a deep notch in its posterior border not found in normal specimens: this peculiarity led to some errors in the original account subsequently corrected by the author [†]. Later descriptions were founded on a nearly perfect shoulder-girdle, which has been figured by Prof. Seeley [‡] and forms the basis of the present account. The closely similar shoulder-girdle of *Baptanodon discus* has been figured and described by Gilmore [§].

The *coracoids* (text-fig. 32) are broad plates of bone thickening greatly towards their inner edge and to a less degree towards the outer edge. The inner thickening (text-fig. 32, C) terminates in a rugose oval surface, placed somewhat obliquely, for union with the opposite member of the pair, there being a pad of cartilage between the two, except perhaps in very old individuals. The upper (visceral) surface is concave from side to side, but the anterior half, owing to the thickening of that region, is convex from before backwards, while the comparatively thin posterior portion is concave in the same direction. The ventral surface is, speaking generally, gently convex from before backwards and concave from side to side. The thin posterior border is evenly convex and in life was fringed with cartilage, as is shown by its irregularly grooved edge. The outer border is occupied mainly by the large, nearly flat glenoid surface, which in life was no doubt covered with cartilage. It widens out from behind forwards, and anteriorly is coterminous with the facet for union with the scapula: this surface makes an angle of about 135° with the glenoid facet; it is nearly triangular and was covered with cartilage. The outer half of the anterior coracoid border, immediately internal to the scapular facet, is thin and sharp-edged; it is deeply concave, forming a sharply-defined notch or bay. This notch seems to represent the remains of the sharp anterior border of the originally much more

---

[*] Quart. Journ. Geol. Soc. vol. xxx. (1874) p. 696, pls. xlv., xlvi. figs. 1, 2; also Proc. Roy. Soc. vol. liv. (1893-4) p. 149.

[†] Proc. Roy. Soc. *tom. cit.* p. 151.

[‡] *Loc. cit.*

[§] "Osteology of *Baptanodon*," Mem. Carnegie Museum, vol. ii. pp. 108–113, text-figs. 21–22, pl. xii.

elongate coracoid, such as occurs in the more primitive members of the group—*e. g.* in *Shastasaurus osmonti*, from the Trias of Northern California, the coracoid of which is figured by Merriam*. The present condition of this notch is due to the great expansion of the bone generally, and there seems to be no sufficient evidence for the view that it formed the posterior border of a foramen closed in front by a precoracoid

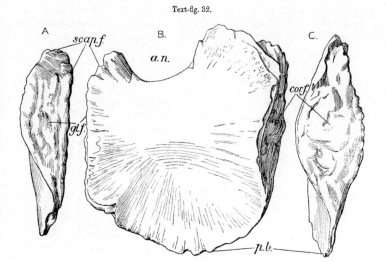

Text-fig. 32.

Left coracoid of *Ophthalmosaurus*: A, outer end; B, upper surface; C, inner (symphysial) end. (R. 2137, ½ nat. size.)

a.n., anterior notch; *cor.f.*, surface for union with opposite coracoid; *gl.f.*, glenoid facet; *p.b.*, posterior border; *scap.f.*, facet for scapula.

cartilage; certainly, even in the earliest forms there is no trace of any ossified precoracoid. No doubt the short anterior prolongation of the coracoid internal to the notch was capped with cartilage, as also was the inner process of the scapula, but there is no evidence that these two cartilages ever joined to enclose a fenestra.

The *scapula* (text-fig. 33) is greatly expanded towards its lower end, which is also greatly thickened posteriorly, while anteriorly it is thinner, and towards its anterior

* Bull. Dept. Geol. Univ. California, vol. iii. (1902-4) pl. x. See also coracoid of *Shastasaurus alexandræ*, figured *op. cit.* pl. xii..

border bent downwards. The thickened portion terminates in a broad cartilage-covered surface, the widest posterior end of which forms the front of the glenoid cavity ($gl.f.$), while in front of this is the articular surface ($cor.f.$) for union with the coracoid. In front of this again the free lower border of the bone may either have been bordered with cartilage, or in old individuals may become smooth and rounded, forming, according to Professor Seeley's interpretation, the outer edge of a fenestra, the other sides of which are constituted by the notch of the coracoid and a precoracoidal cartilage. In front of this, as far as the antero-internal angle, the inner border is again thickened and deeply grooved and pitted for union with the

Text-fig. 33.

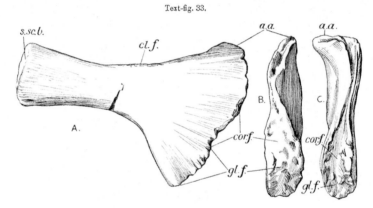

Scapula of *Ophthalmosaurus*: A, from above; B, proximal end; C, proximal end of scapula of the type specimen. (A & B, R. 2137; C, R. 2133: about ½ nat. size.)

a.a., acromial angle; *cl.f.*, facet for clavicle; *cor.f.*, facet for union with coracoid; *gl.f.*, glenoid surface; *s.sc.b.*, suprascapular border.

cartilage which, according to the views just referred to, was continuous with that capping the anterior border of the inner side of the coracoid and was homologous with a precoracoid element. The upper (visceral) face of the expanded portion of the scapula is flat or slightly concave and is deflected sharply downwards in front, as mentioned above; the degree of deflection of this anterior border differs very considerably in different specimens, but in no case is a broad flat anterior surface formed for union with the clavicle, such as is figured by Gilmore (*op. cit.*) in the shoulder-girdle of *Baptanodon*; there may, however, be an extension of the rough clavicular surface, best marked on the anterior edge of the blade of the scapula, so that the union with the clavicle may have extended to this deflected portion, which Professor Seeley regards as equivalent to the acromion process. The ventral surface is concave from before backwards. The blade is narrow and is slightly curved upwards; it is nearly the same width throughout, widening only a little towards its upper end, which is truncated nearly at right angles by a surface which seems to have borne a cartilaginous lip or suprascapular cartilage in life. The anterior, slightly concave border of the blade is rounded from above downwards, and the lower portion of its inferior edge bears a long roughened surface for union with the clavicles; in some cases the roughened area at this point is so marked that it forms a slight projection, which might perhaps be regarded as a sort of acromion process. The posterior border of the blade is also roughened, presumably for the insertion of muscles.

The clavicular arch (text-fig. 34) consists of a pair of clavicles and a T-shaped interclavicle. The *clavicles* (text-fig. 34, A, B, C) are long curved elements terminating externally in a point and internally in a complex and irregular sutural border, by which they interdigitate and unite more or less firmly in the middle line, being further keyed together by the anterior bar of the interclavicle. The external portion is more or less oval in section and bears on its posterior side a roughened surface for ligamentous union with the corresponding surface on the front of the scapula above referred to. Internal to this outer horn-like portion the bone widens out and consists of a thickened anterior rim from which a thinner shelf-like region projects upwards and backwards almost at right angles, the two forming the deeply grooved inner surface to the bone (text-fig. 34, C, *i.cl.g.*). It is into this groove towards the median portion of the bone that the arm of the T-shaped interclavicle fits, the union in some cases being extremely close and perhaps in old individuals in places amounting to actual fusion of the two elements. The antero-dorsal surface of the clavicle is smoothly convex from before backwards in its outer portion, but towards the middle it bears several strong ridges which terminate in the digitations of the median sutural surface.

The *interclavicle* (text-fig. 34, D, E, F) consists of an anterior transverse bar and a posterior median process. The anterior portion thins away towards each end, terminating in a blunt point. Dorsally it is deeply grooved like the clavicles, its anterior nearly flat face making about a right angle with the rounded ventral portion, which is raised in the middle into a roughened boss of bone: this remains uncovered by the clavicles, which embrace the remainder of the ventral and the lower portion of the anterior face of the transverse bar (text-fig. 34, D, F). The posterior bar has its ventral face convex from side to side, while the dorsal is convex anteriorly, but posteriorly bears an elongated flattened facet (*cor.f.*) which must have been applied to the ventral surface of the intercoracoidal cartilage, or perhaps in old individuals to the ventral surface of the coracoids themselves.

The *humerus* (text-figs. 36 & 37) is a short but very massive bone, the upper end being particularly solid, while the lower is compressed in the plane of the paddle.

Text-fig. 34.

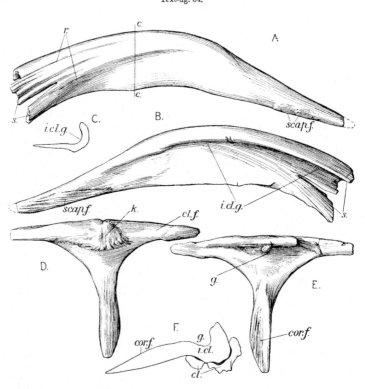

Clavicle and interclavicle of *Ophthalmosaurus*: A, right clavicle from front; B, ditto from behind; C, section across A at the line c–c; D, interclavicle from below; E, from above; F, section across clavicle and interclavicle along the middle line of the latter. (R. 2137, about ½ nat. size.)

c–c, line of section drawn in fig. C; cl., clavicle; cl.f., facet for clavicle; cor.f., facet for coracoid or intercoracoidal cartilage; g., groove in upper surface of transverse bar of interclavicle; i.cl., interclavicle; i.cl.g., groove in upper surface of clavicle for the reception of the anterior bar of the interclavicle; k., knob on ventral face of interclavicle; r., ridges on anterior face of clavicle; s., interdigitating suture of clavicles; scap.f., facet for anterior border of scapula.

The description of the humerus is rendered difficult by the circumstance, that all the specimens in the collection are free and not in their natural position in relation to the rest of the skeleton, as they usually are in the specimens of the Liassic Ichthyosaurs. Consequently it has not been easy to determine with certainty which is the right and which the left, and therefore also which is the dorsal and which the ventral (palmar) face. A further difficulty is also introduced by the fact that when these animals died the paddles seem sometimes to have fallen over on to their dorsal surface, even when the remainder of the skeleton clearly rested on its ventral face. The position of the paddles in several specimens has been carefully noted by Mr. A. N. Leeds, and the following determination of the right and left, and of the upper and lower face,

Text-fig. 35.

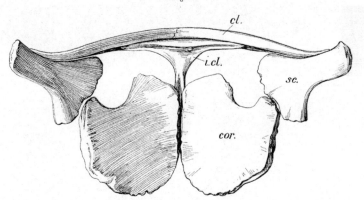

Upper surface of restored shoulder-girdle of *Ophthalmosaurus*. (R. 2137, about ¼ nat. size.)
cl., clavicle ; *cor.*, coracoid ; *i.cl.*, interclavicle ; *sc.*, scapula.

depends mainly on his observations. It should be noted that the account of this bone and of the femur given in the 'Geological Magazine' (dec. 5, vol. iv., May 1907, p. 202 *) is erroneous so far as these points are concerned.

In the following description the limb is supposed to be in such a position that the expanded paddle has its palmar (ventral) surface looking downwards and backwards, and making an angle of about 45° with the vertical.

The direction of the paddle with reference to the proximal end of the humerus is marked in text-fig. 36 by the line a–b, a being uppermost. In this case the border

* See also Geol. Mag. [5] vol. v. p. 96.

of the proximal end which is marked *l.b.* in the figure, will be nearly vertical, and *p.b.* will form the upper edge, which is somewhat curved and makes an angle a little

Text-fig. 36.

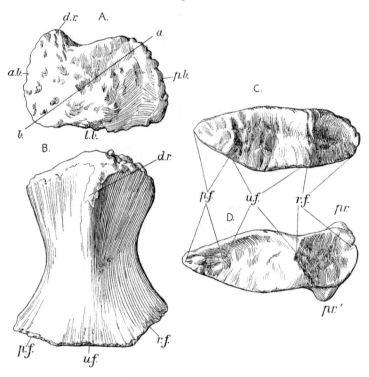

Left humerus of *Ophthalmosaurus*: A, proximal end; B, ventral (palmar) side; C, distal end; D, distal end of another example showing the preaxial prominences. (A–C, R. 2134; D, R. 2160: all ⅓ nat. size.)

a.–b., line marking the direction of the distal expansion of the bone with reference to the proximal end; *a.b.*, lower border; *d.r.*, deltoid ridge; *l.b.*, anterior border; *p.b.*, upper border; *p.f.*, facet for pisiform; *pr., pr'.*, projections on anterior angle; *r.f.*, facet for radius; *u.f.*, facet for ulna.

less than a right angle with *l.b.* Posteriorly it passes by a rounded angle into the posterior border, which forms a sigmoid curve, the lower end rising into the anterior edge of a strong deltoid crest (*d.r.*), which extends somewhat obliquely down the shaft to a point a little above the middle. The lower border (*a.b.*) is somewhat convex and makes a right angle with the anterior face (*l.b.*). The whole of the proximal end of the bone is greatly roughened by a series of irregular rounded prominences, clearly indicating that in life there was a thick cap of cartilage which seems also to have extended on to the summit of the deltoid crest. The appearance of this cartilage-covered region is quite unlike the nearly smooth and fibrous-looking surface of the shaft.

The shaft is short and stout and is considerably contracted in the middle, at the same time becoming compressed dorso-ventrally; its preaxial border is a sharp angle, while posteriorly it is more rounded. Distally the bone expands greatly, the long axis of the expansion making an angle of about 45° with the longest axis of the proximal end, as already described. The effect of this peculiarity is to give the bone the appearance of having undergone slight torsion.

In many specimens the ventral surface of the distal expansion bears near its preaxial angle a prominent tubercle (*pr.*), while on the dorsal surface nearly opposite there is sometimes an even more prominent projection (*pr.'*) which may form a blunt proximally-directed process. These projections seem to have served for the insertion of the tendons of muscles; but it is remarkable that in many large and apparently quite adult specimens they are entirely absent, and not the smallest trace of muscle-attachments where these processes should be can be detected. It is difficult to suppose that this difference in the musculature can be a specific character, and it seems more probable that the presence or absence of the powerful muscles indicated by these processes was a secondary sexual character, a conclusion that is to some degree supported by the occurrence of about equal numbers of the two forms. The distal articular surface bears three distinct facets, each concave and covered with rugosities indicating the presence in life of a covering of cartilage. The anterior or radial surface (*r.f.*) is smaller than that for the ulna and makes an angle of about 135° with it. The ulnar surface (*u.f.*) is large and roughly quadrate in form, its upper and lower borders being convex, the anterior and posterior being formed by slight ridges separating it from the anterior (radial) and posterior (pisiform) facet (*p.f.*) respectively. The posterior facet is much smaller than the others and sometimes not very distinctly marked off from the ulnar surface; it is triangular in outline and articulates with the bone which is here called the pisiform, that element of the proximal row of the carpus having apparently shifted back till it acquired a surface of union with the humerus. In some specimens the posterior facet is so small and slightly marked that it is clear that in these cases the pisiform was only just in contact with the humerus. The articulation of this bone with the humerus is no doubt correlated with the increasing breadth of the paddle, but the same end seems to have been attained in other genera by different means. Thus in the fore paddle described by Mr. Boulenger * as the type of *Ichthyosaurus extremus* it is clear that the third bone is the intermedium which is thrust between the radius and ulna, and articulates extensively with the humerus; this same condition is known in other species of *Ichthyosaurus* and appears in some cases to occur also in the hind paddle †.

Beneath the humerus the remaining bones of the paddle form a mosaic (text-fig. 37), the proximal rows having been separated by comparatively thin layers of cartilage, while the more distal phalanges seem to have been embedded in a mass of that substance and separated from one another by considerable intervals. The *radius* (text-fig. 37, *r*.) is a pentangular or subquadrate bone, the anterior (outer) border of which is thin and sharp and continues the line of the radial border of the humerus; internally the bone thickens greatly. Proximally it unites with the humerus and internally (postaxially) it usually touches the ulna, but the length of its contact with that bone varies greatly in different individuals according to the degree to which the intermedium is thrust between them. Postero-internally and distally it has long surfaces for the intermedium (*int*.) and the radiale (*rad*.) respectively. The *ulna* (*u*.) is a large subquadrate bone articulating with the humerus, usually to a greater or less degree with the radius, as already mentioned, with the intermedium and ulnare, and, finally, posteriorly with the *pisiform* (*p*.). This, the third element articulating with the humerus, is of a somewhat doubtful character, though the interpretation here followed is now usually adopted. Prof. Seeley, in his original paper on *Ophthalmosaurus*, called it the olecranon, and recently Prof. Williston has proposed for it the non-committal name—epipodial supernumerary. This element is oval, and its posterior border is thin, though it seems to have continued fringed with cartilage throughout life: the extent of its articulation with the humerus, as above noted, varies much, but in all cases is comparatively small. It may be stated at once that distally the pisiform articulates with a smaller oval element, which in its turn bears a small distal nodular bone (in some cases two), the whole apparently representing a rudimentary fifth digit. Leaving the pisiform and its appendages out of account, the *manus* may be described as consisting of a carpus (the proximal row of which consists of three bones, the distal of four), and four digits composed of a varying number of bones. A certain degree of variation in the arrangement of the carpal bones occurs, but that here described seems to be the normal one. Unfortunately in the paddle figured the second carpal of the distal row is much smaller than usual and presents somewhat the appearance of a centrale articulating between the intermedium and radiale, but its true character is

---

* Proc. Zool. Soc. 1904, vol. i. p. 424. The horizon and locality from which Mr. Boulenger's specimen came are not known, and he regarded the species as probably Liassic; but a humerus of precisely similar type was lately obtained from the Kimmeridge Clay of Swindon by Mr. Arthur D. Passmore, so that it is likely that the type specimen is also from the later horizon, as from its extreme specialization would be probable.

† See Lydekker, Catal. Foss. Rept. Brit. Mus. pt. ii. p. 43, text-fig. 20. Also Fraas, 'Die Ichthyosaurier d. Süddeutschen Trias- und Jura-Ablagerungen,' pl. vi. fig. 4.

OPHTHALMOSAURUS.

clear on comparison with other specimens. In the proximal row of the carpus the largest element is the *intermedium* (*int.*): this bone thrusts itself to a varying degree

Text-fig. 37.

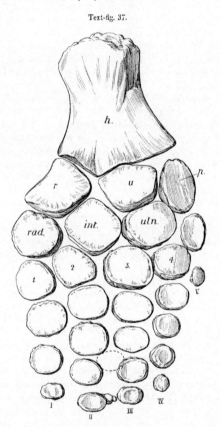

Dorsal view of left fore paddle of *Ophthalmosaurus*. (R. 2853, $\frac{1}{4}$ nat. size.)
h., humerus ; *int.*, intermedium ; *p.*, pisiform ; *r.*, radius ; *rad.*, radiale ; *u.*, ulna ; *uln.*, ulnare ;
1–4, the distal row of carpals ; I–V, the number of the digits.

between the radius and ulna; laterally it bears short facets for union with the radiale and ulnare, and distally it unites with the second and third elements of the distal row.

The *radiale* (*rad.*) is a rounded, somewhat pentangular bone, thinning towards its anterior edge, which, however, is not sharp as in the radius, but was fringed with cartilage throughout life; it articulates with the radius, intermedium, and the first and second of the distal carpals. The *ulnare* (*uln.*) is a roughly quadrate element articulating with the ulna, intermedium, and the third and fourth of the distal carpals; it may also have been in contact with the pisiform. No centrale can be recognized. The four distal carpals are more or less oval or rounded in outline; the two middle elements articulate with the intermedium. The remaining bones of the paddle are arranged in four rows, at least in most specimens, the more proximal being oval in outline, with the long axis transverse, while the more distal elements may be merely small rounded masses of bone. No specimen has been collected in which these bones have been retained in the matrix in their natural positions, so that there is some doubt as to their precise arrangement. In one of the best specimens carefully mounted by Mr. Leeds as found in the rock, the first digit consists of three elements, the second, third, and fourth of four each; but, judging from the other examples, the phalanges may have been more numerous, the terminals being represented by mere nodules of bone easily overlooked in removal.

Another disadvantage of not having the paddles actually embedded in the matrix is that it is impossible to say whether any of the digits ever bifurcated, as sometimes happens in the Ichthyosaurs, or whether there were accessory ossicles along the margins of the paddles: in one example it seems possible that there was a small radial sesamoid which may have been in contact with the humerus and perhaps formed the proximal end of a rudimentary additional digit on the preaxial side of the fin. From examination of the best-preserved paddles, it seems that there is no reason for regarding the digits present as other than those of the primitive pentadactyl limb, and the type of structure might easily be derived from such a paddle as that of the Triassic *Mixosaurus cornalianus* as figured by Merriam [*] by way of some latipinnate Liassic *Ichthyosaurus*, as, in fact, has been suggested by Merriam [†].

Pelvic Girdle and Hind Limb.—The pelvis has undergone great reduction, considerably greater than in the Liassic Ichthyosaurs. It is improbable that the ilium articulated above with the sacral ribs, and the pubes and ischium are fused with one another at their proximal and distal ends, their original separation being usually marked only by a narrow slit-like foramen and occasionally a notch between their ventral ends.

The *ilium* (text-fig. 38) is a curved sickle-like bone laterally compressed into a thin blade above, but thickening towards the acetabular end, where it terminates in a cartilage-covered surface, which in some specimens is distinctly divided into two portions, which make an angle with one another, one being the surface for union with

---

[*] "Triassic Ichthyosauria," Amer. Journ. Sci [4] vol. xix. (1905) p. 25, fig. 1.

[†] *Loc. cit.* p. 27.

OPHTHALMOSAURUS.

the pubis and ischium (*f.*), the other, the larger of the two, forming the upper part of the acetabulum (*acet.*). The inner face (*i.s.*) of the upper end of the blade bears an elongated roughened surface, which may indicate the persistence of some loose ligamentous connection with the vertebral column. The pubis and ischium (text-figs. 39, 40), as already noticed, are fused together, and, as might be expected in bones undergoing reduction, they differ in form to a great extent in different individuals. The upper end of the fused bones is thickened considerably and terminates in cartilage-capped surfaces, the smaller of which is on the inner side and forms the articulation for the ilium (*il.f.*, text-fig. 40); the larger occupies the greater

Text-fig. 38.

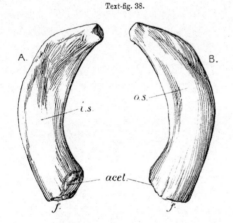

Right ilium of *Ophthalmosaurus* (R. 2853, ⅔ nat. size): A, inner side; B, outer side.
acet., acetabulum; *f.*, facet for union with the ischio-pubis; *i.s.*, inner surface; *o.s.*, outer surface.

part of the proximal ends of the combined bones and formed part of the acetabulum (*acet.*). Ventrally the bones become greatly flattened from within out, so that at their lower end they form a thin plate of bone. About the middle of their length occurs the slit-like obturator foramen (*ob.f.*), a persistent portion of the original separation of the two elements, which also is sometimes indicated by a notch at their distal ends. The *ischium* (*isch.*) is by far the larger of the two bones, forming a comparatively broad plate at its ventral end. The lower border is convex and it is doubtful whether the bone actually met its fellow in a median symphysis. The *pubis* (*pu.*) is relatively narrow, widening out a little ventrally. The fusion of the two bones at their upper end is so complete that no trace of their original separation is visible, and consequently the share of each in the articular surface for the ilium and in the acetabulum cannot be made out.

The *femur* (text-fig. 41) is a very much smaller bone than the humerus, the relative lengths of the two being about as 2 to 3, while the difference in bulk is still more striking. The proximal end of the bone is very massively constructed compared with

Text-fig. 39.

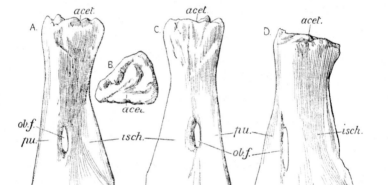

Left ischio-pubes of *Ophthalmosaurus*: A, from outer side; B, proximal end; C, inner side; D, a second specimen from outer side (A, B, C, R. 2853, ⅔ nat. size).

acet., acetabulum; *isch.*, ischium; *ob.f.*, obturator foramen; *pu.*, pubis.

the distal end, and it agrees generally in form with the proximal end of the humerus. If, as in the description of the fore limb, we assume that the palmar face of the paddle looks downwards and backwards, making an angle of about 45° with the vertical, then in text-fig. 41, C, which represents the proximal end of the right femur, the line a–b shows the direction of the paddle, a being at the top. The border marked *a.b.* is ventral, while *l.b.* is anterior. The large trochanter-like ridge *d.r.* forms the postero-inferior angle on the border, *p.b.* is at the top. As in the case of the humerus, the upper end of the bone, including the upper part of the trochanteric ridge, is covered by coarse rugosities, showing that there was an external cap of cartilage. From the head the shaft narrows rapidly, at the same time becoming compressed dorso-ventrally towards the distal expansion, which is relatively smaller than in the humerus. Distally the femur articulates with two bones only, the tibia (*t.*) and fibula (*f.*), the facets for which are shown in text-fig. 41, D (*t.f.* & *f.f.*). It will be seen in the specimen figured

Text-fig. 40.

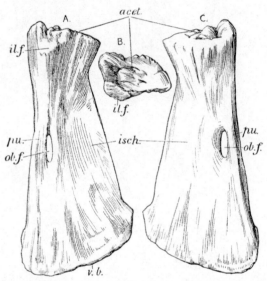

Left ischio-pubis of *Ophthalmosaurus* (⅔ nat. size): A, from outer side; B, proximal end; C, from inner side. *acet.*, acetabulum; *il.f.*, facet for ilium; *isch.*, ischium; *ob.f.*, obturator foramen; *pu.*, pubis; *v.b.*, ventral border.

that the smaller of these two bones is the tibia and that it is irregularly oval in outline, articulating proximally with the femur, distally with the tibiale and intermedium, and postaxially with the fibula.

In some other specimens the ossification of the anterior edge of the tibia is more extensive, so that the anterior border is sharp and slightly concave, and the whole bone as large as or larger than the fibula. This latter is roughly pentangular in shape; it

Text-fig. 41.

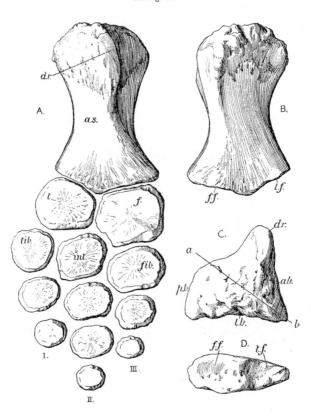

Right hind limb of *Ophthalmosaurus* (⅔ nat. size): A, ventral face of right hind paddle; B, dorsal side of right femur; C, proximal end of right femur; D, distal end of right femur.

a.-b., line marking the direction of the long axis of the paddle in relation to the proximal end of the femur; *a.b.*, ventral border; *a.s.*, ventral surface; *d.r.*, trochanteric ridge; *f.*, fibula; *f.f.*, fibular facet; *fib.*, fibulare; *int.*, intermedium; *l.b.*, anterior border; *p.b.*, dorsal border; *t.*, tibia; *tib.*, tibiale; I–III, the number of the digits.

articulates with the femur proximally, the tibia anteriorly, and the intermedium and fibulare distally. Of the proximal row of carpals, the tibiale (*tib.*) is the smallest, the intermedium (*int.*) somewhat larger, and the fibulare (*fib.*) larger still, the increase in dimensions being mainly from side to side. As usual, the intermedium (*int.*) joins both the tibia and fibula, its union with the former being the most extensive. The distal row of tarsals also consists of three oval elements, the middle one being the larger. In the most satisfactorily preserved hind paddle (text-fig. 41) there is a row of three metatarsals; of these the middle one alone has a phalangeal ossicle articulating with it distally: probably there were other small nodules of bone representing phalanges which have been lost. It is clear that the hind paddle consisted of three digits (I.–III.) and has undergone extensive reduction.

Ophthalmosaurus icenicus, Seeley.

[Plate I.; text-figs. 1–42.]

1874. *Ophthalmosaurus icenicus*, H. G. Seeley, Quart. Journ. Geol. Soc. vol. xxx. pp. 696–707, pls. xlv. & xlvi.
1889.   ,,   ,,   R. Lydekker, Catal. Foss. Rept. British Museum, pt. ii. p. 9.
1890.   ,,   ,,   R. Lydekker, *op. cit.* pt. iv. p. 267, fig. 62.
1905–6.   ,,   ,,   C. W. Gilmore, Mem. Carnegie Museum, vol. ii. pp. 125, 336.

Type Specimen.—An imperfect skeleton (R. 2133) in the British Museum.

This species was originally founded by Professor Seeley on a shoulder-girdle, which occurred associated with the incomplete skull and skeleton of a single large individual (Pl. I. figs. 11–15; text-figs. 4, 26, 33 C).

In the present Catalogue it has been found impossible to distinguish more than a single species. If only a few skeletons had been preserved several forms would probably have been recognised and named, since taken individually some specimens seem to differ very considerably from others. Moreover, if a single element of the skeleton, say the quadrate, were taken as a means of determining the species, several probably might have been distinguished, but these certainly would not correspond with the specific divisions that might be founded on other elements. In examining the immense collection made by Mr. Leeds and only in part catalogued below, it becomes clear that these Ichthyosaurs vary considerably in details of the structure of the various parts of the skeleton. The reasons for this extreme variability seem to be, (1) differences in the form of the bones resulting from the varying degree to which ossification has proceeded in a skeleton in which throughout life much cartilage persisted; (2) the mode of preservation of the bones and the degree to which they have been distorted by earth-pressure; (3) differences in the age and sex of the individuals. It has therefore been thought best to refer all the

Text-fig. 42.

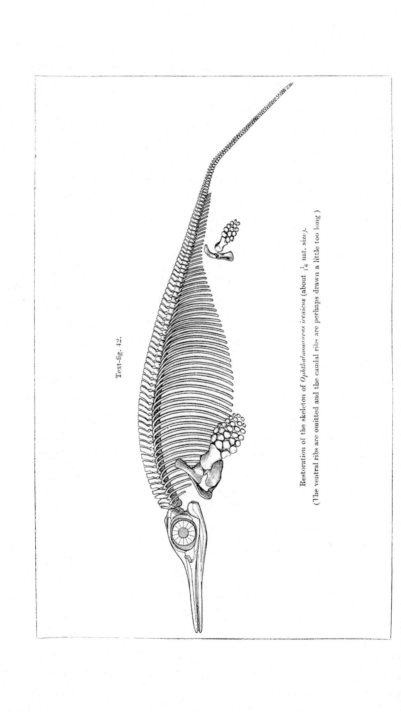

Restoration of the skeleton of *Ophthalmosaurus icenicus* (about $\frac{1}{16}$ nat. size).

(The ventral ribs are omitted and the caudal ribs are perhaps drawn a little too long.)

specimens to a single species, while at the same time admitting that future investigation may render possible the diagnosis of others.

Form. & Loc.—Oxford Clay: England.

All the following specimens from the Leeds Collection were obtained from the Oxford Clay in the neighbourhood of Peterborough.

R. 2133 (Leeds Coll. 62). Imperfect skeleton, including the shoulder-girdle described and figured by Prof. Seeley in the Quart. Journ. Geol. Soc. vol. xxx. (1874) pp. 696–707, pls. xlv. & xlvi. figs. 1–2, as the type of the genus and species. The portions of the skeleton preserved include : the basioccipital (Pl. I. figs. 13, 14), both quadrates (Pl. I. fig. 11) with portion of the squamosal attached, right and left stapes (Pl. I. fig. 12), left opisthotic (Pl. I. fig. 15), left parietal, and several other fragments of the skull ; some sclerotic plates ; the posterior portion of the left ramus of the mandible ; a considerable portion of the vertebral column (text-fig. 26), including the atlas, axis, and thirty-five other precaudal vertebræ, together with twenty-one caudals mostly from the anterior part of the tail; a number of separated and imperfect neural arches and many fragments of ribs ; the shoulder-girdle, of which only the right scapula (figured *loc. cit. supra*, pl. xlv. fig. 2) and part of the clavicular arch (pl. xlvi. figs. 1–2) are of normal shape, the other bones being deformed ; fourteen carpals and phalanges. The left scapula is an almost shapeless mass of irregular bone fused with the upper end of the clavicle, and the right coracoid (figured *loc. cit. supra*, pl. xlv. fig. 1) has a deep notch on its posterior border, which is not present in normal specimens.

Dimensions (in centimetres) :—

Basioccipital: total length 7·5 ; length of neural surface 5·1 ; greatest width 9 ; width of condyle 6·7 ; depth of condyle 5·8.

In this specimen the central portion of the condyle is much more convex than the lateral portion.

Quadrate: extreme height 13 ; width of neck 6·3 app. ; width of articular surface 5·2 ; length of ditto 7·6.

Vertebræ.	Atlas and axis.	Cervical (about the 5th).	Posterior dorsal.	First caudal.
Height of centrum	7·2 (?)	6·9	8·2	8·2
Width of centrum	7·4	7·3	9·8	9·6
Length of centrum (measured at neural canal)	4·5	3·8	4·4	4·1
Greatest width with diapophyses	8·3	8·6	10·5	9·8

The total length of the precaudal portion of the column, as preserved in this specimen, is about 1·60 metres, but probably some vertebræ are wanting.

Scapula: greatest length . . . . 25·0
width of proximal expansion . . . . 16·5
blade (at narrowest) . . . . 4·6
Coracoid: width . . . . 20·7
length . . . . 21·8

R. 2180 (Leeds Coll. 76). The greater part of the skull and skeleton. The bones of the skull are for the most part separated from each other and several are figured in the text, as noted below. The portions preserved include: basioccipital, exoccipitals, basisphenoid with parasphenoid attached (text-fig. 7), opisthotics, stapes, quadrates, quadrato-jugals (text-fig. 10), jugals (text-fig. 10), maxillæ (text-fig. 13), premaxillæ, nasals (text-fig. 12), lachrymals (text-fig. 11), postorbitals (text-fig. 8), part of prefrontal, postfrontals, together with fragments of squamosals and parietals and other roofing-bones of the skull. The pterygoids (text-fig. 17), one palatine, and perhaps the vomers are preserved. The mandible (text-figs. 20 & 22) is nearly complete, only its tip being missing, and the bones of the right ramus are mostly separate from each other. In this skull and mandible many of the bones are crushed and some wanting, but the circumstance that they are for the most part disarticulated makes the specimen especially valuable, and it has been employed in the text in the description of many of the individual elements and as the basis for the restoration given in text-fig. 23. A number of sclerotic plates, mostly separated from each other and incomplete at their outer ends, are preserved with this skull. There are also about half a dozen teeth, with slightly fluted, pointed crowns and rather deeply grooved roots. The rest of the skeleton is represented by the atlas, axis, thirty-two other precaudal vertebræ, and twenty-four caudals: many of these vertebral centra are much crushed and distorted and none remain united to the neural arches. About thirty more or less imperfect and crushed neural arches are preserved: of these about six, probably belonging to the anterior cervicals, seem to be composed of two halves (neurapophyses), which united in suture with each other in front and overlapped the anterior part of the next neural spine behind (text-fig. 29). There are numerous double-headed precaudal ribs and a few single-headed caudal ribs, mostly more or less imperfect.

The shoulder-girdle is complete, though somewhat crushed. Both humeri are preserved, also the radius, ulna, pisiform, radiale, intermedium, ulnare, and about thirty other paddle-bones.

The pelvis and hind limbs are wanting.

Some approximate measurements (in centimetres) of some parts of this skeleton are given below:—

Basioccipital: greatest width	7·9
greatest length	6·5
length of neural surface	4·0
width of occipital condyle	5·9
height of occipital condyle	4·6
Basisphenoid and parasphenoid (text-fig. 7): total length	30·0
Parasphenoid: length of free portion	23·5
width at base	2·0
Jugal (text-fig. 10): total length	26·5
width of suborbital bar	2·0
Lachrymal (text-fig. 11): length of base	12·5
height	6·4

Maxilla (text-fig. 13): length . . . . . . (approx.)	26·0
greatest depth	5·2
width at palatine plate	2·8
Premaxilla: length in front of anterior angle of nares . . .	49·0
depth . . . . (approx.)	4·6
Pterygoid (text-fig. 17): length . . . . . . (approx.)	43·0
Quadrate: greatest length	13·4
width of neck	5·5
,, articular surface	5·1
Articular bone (text-fig. 22): length of articular surface . .	7·1
length	5·9
anterior depth	4·9
Atlas and axis (crushed): length	4·3
Anterior precaudal (? cervical): length	2·8
width	6·2
Middle precaudal: length	3·8
width	8·0
Anterior caudal: length	2·5
width	7·8
Anterior caudal (crushed, see text-fig. 31, B): length . . .	1·7
width	8·0
Middle caudal (near bend): length (at neural canal)	1·9
width	4·4
Posterior caudal (behind bend): length	1·7
width	2·7
Anterior (cervical) neural arch (text-fig. 29): height to top of spine	8·2
Posterior (dorsal) neural arch: height to top of spine . . .	11·3
antero-posterior width of spine.	3·0
Clavicular arch: length of clavicle (in straight line)	30·0
width between outer ends of clavicles . . .	48·0
length of interclavicle in mid-line	12·9
Scapula: greatest length	20·0
width of proximal expansion	13·2
,, blade (in middle) . . . . . (approx.)	4·4
Coracoid: approximate width	19·0
,, greatest length	21·0
Humerus: length	15·5
width of shaft at narrowest	7·1
,, distal end	11·5

R. 2181 (Leeds Coll. 66). Imperfect skull and skeleton of a very young individual. In this specimen the anterior portions of the jaws bear a number of small, very closely crowded and rather backwardly directed teeth (Pl. I. figs. 7, 8): in the anterior part of the right side of the mandible there are ten on a space of 5·5 cm. The pubis and ischium show no more sign of separation than in older individuals, and the same may be said of the atlas and axis vertebræ. The portions of the skull preserved include: basioccipital, stapes (1), opisthotics, squamosal, jugals, lachrymal, nasals, quadrates, pterygoid; also a considerable part of the mandible, including both articular bones, and about forty teeth; hyoid bones.

Of the skeleton there are: atlas and axis and sixty-five other vertebræ, a few neural arches, numerous fragments of ribs, right clavicle, coracoids, scapulæ, humeri, radius, ulna, ilia, parts of ischio-pubes, femora.

The dimensions (in centimetres) of some of the bones are:—

Basioccipital: width of condyle	3·2
Quadrate: extreme length	6·8
width at neck	3·3
length of articulation (approx.)	3·5
Articular bone: length	3·3
Hyoid bone: length	9·2
width of middle of shaft	0·7
Atlas and axis: length	1·9
greatest width	4·6
Middle precaudal: width	4·5
Anterior caudal: length	1·4
width	1·9
Coracoid: greatest width	8·6
Scapula: greatest length	10·2
width of proximal expansion	6·6
Humerus: length	7·1
width of distal end (approx.)	5·4
Ilium: length	5·3
Ischio-pubis: length (approx.)	6·6
Femur: length	4·1
width of distal end	2·6

R. 2740. A large part of the skull and mandible of a large individual. The bones are well preserved but greatly displaced, and in part imbedded in hard matrix: a portion of the sclerotic ring is preserved *in situ*, and the cut edges of the plates show that they are closely interlocked along their edges (Pl. I. fig. 10). The chief parts shown are: basioccipital with exoccipitals and supraoccipital attached, basisphenoid with the parasphenoid, postero-superior portion of orbit showing the relations of the postorbitals, supratemporal, the anterior wing of the squamosal, and the postfrontal to one another; the facial region with the nasals, lachrymals, maxillæ, and premaxillæ nearly in their natural relations. The mandible is represented by the greater part of the postsymphysial region of both rami.

Some dimensions (in centimetres) of this specimen are:—

Basioccipital: width of condyle	7·7
greatest width	10·9
,, length	8·5

Basispheoid: width . . . . 11·0
      length . . . . 8·7
Parasphenoid: length of free portion . . . . 19·0+
Lachrymal: length of base . . . . (approx.) 12·0
Premaxilla: length in front of anterior angle of nares . . . 52·0
Articular bone: length . . . . 6·0
Surangular (text-fig. 21): length . . . . . . (approx.) 57·0

R. 2160 (Leeds Coll. 65). Portions of the skull including the basioccipital, exoccipitals, supra-occipital, stapes, opisthotics, one pro-otic, quadrates, basisphenoid, portions of the roof and snout; the articulars and other portions of the mandible, some sclerotic plates; a large part of the vertebral centra including those of the atlas and axis, and of about twenty-four other precaudal and sixty caudal vertebræ; three neural arches of posterior caudal vertebræ; scapulæ, coracoids, humeri, radii, ulnæ, pisiforms, carpals, and about sixty other bones of the front paddles; one imperfect ischio-pubis, femora, tibiæ, fibulæ, and seven other bones of the hind paddles. This specimen differs in several respects from the usual type and, if a second species is at any time established, it would certainly have the best claims to become the type of a new form. The most striking peculiarity is to be found in the quadrates, in which the posterior border above the articulation instead of being merely concave is deeply notched; on the basioccipital the occipital condyle is somewhat less convex and more prolonged ventrally than in most specimens, though this form is found in some skeletons with the ordinary type of quadrate. The coracoids are perhaps a little longer in proportion to their width, and the anterior notch rather smaller than is the case with the ordinary forms. In the humeri the prominences near the anterior angle of the distal end are well developed on both upper and lower surfaces (see text-fig. 36, D, $pr.$, $pr.'$).

The dimensions (in centimetres) of some parts of this skeleton are:—

Basioccipital: greatest width . . . . 9·2
    ,, length . . . . 6·1
    width of occipital condyle . . . . 7·1
    height of occipital condyle . . . . 7·0
Stapes: length . . . . 7·2
Quadrate: extreme length . . . . 11·9
    width of neck . . . . 5·5
    ,, articular surface . . . . 4·8
Articular bone: length . . . . 5·8
Coracoid: length . . . . 23·3
    width . . . . 18·2
Scapula: length . . . . 21·3
    width of blade . . . . 5·0
    ,, distal expansion . . . . 14·4
Humerus: length . . . . 15·6
    width of shaft at narrowest . . . . 6·9
    ,, distal end . . . . 12·9

Femur: length . . . . 9·8
        width of shaft at narrowest . . . . 3·6
        „   distal end . . . . 6·8
Ischio-pubis: length . . . . 13·3

R. 2853 (Leeds Coll. 85). Fragments of skull, including pterygoids, sclerotic plates, six centra of vertebræ, both scapulæ, fragment of coracoid and clavicles, interclavicle, both fore paddles complete (the left is figured in text-fig. 37, and also by A. S. Woodward in 'Outlines of Vertebrate Palæontology,' p. 182, fig. 113, B), ilia (text-fig. 38), both ischio-pubes (text-fig. 39, A–C), both femora and numerous paddle-bones.

The dimensions (in centimetres) of some of the bones are :—

Scapula: greatest length . . . . 25·0
       width of blade at narrowest . . . . 5·2
       „   distal expansion . . . . 14·6
Right humerus: length . . . . (approx.) 19·0
       width of shaft at narrowest . . . . . . 10·1
       „   distal end . . . . 15·5
Right radius: width . . . . 6·5
       length . . . . (approx.) 6·5
Right ulna: width . . . . (approx.) 8·0
       length . . . . 4·8
Right intermedium: width . . . . 7·8
       length . . . . (approx.) 6·0
Ilium (text-fig. 38): length in straight line . . . . 12·0
       greatest width . . . . 3·1
       width of lower end . . . . 3·2
Ischio-pubis (text-fig. 39, A–C): length . . . . 15·9
       width of proximal end . . . . 4·0
       „   distal end . . . . . 6·9
Femur: length . . . . 11·5
       width of shaft at narrowest . . . . 3·4
       „   distal end . . . . 6·5
Tibia: length . . . . 3·8
       width . . . . 3·7
Fibula: length . . . . 3·2
       width . . . . 3·9

R. 2149 (Leeds Coll. 70). Portions of a skull and mandible, including basisphenoid, stapes, opisthotics, one pro-otic, squamosal, parietal, both articulars, one surangular, and other fragments; atlas and axis, thirty other precaudal and twenty-five caudal vertebræ; the clavicles, scapulæ, coracoids, left humerus, and twenty-three paddle-bones; one ilium and both imperfect ischio-pubic bones. This skeleton does not present any very notable features : the vertebræ are for the most part crushed ; the clavicles show clearly the surfaces by which they unite with the cross-bar of the interclavicle ; the ischio-pubic bones are rather shorter and stouter than in most specimens.

OPHTHALMOSAURUS ICENICUS.

Some dimensions (in centimetres) of parts of this skeleton are:—

Basisphenoid: length	7·2
greatest width	10·5
Quadrate: length	14·5
width at neck	6·9
Articular: length	6·0
Atlas and axis: width	9·5
height	8·4
length (approx.)	3·3
Clavicle: length in straight line (approx.)	36·0
Scapula: length	21·5
width of proximal expansion	14·5
Coracoid: length	23·0
width	19·2
Humerus: length	16·6
width of distal end	13·5
Ischio-pubis: length	13·4
width of proximal end	4·8

R. 2138 (Leeds Coll. 64). Some fragments of a skull, including the basioccipital, basisphenoid, and part of a quadrate, also the articular bones, associated with the atlas and axis, thirty-seven other precaudal and forty-eight caudal vertebræ, three caudal ribs, and two neural arches from the same region; the imperfect scapulæ and coracoids, both humeri with the radii, ulnæ, and pisiforms, as well as forty-two other paddle-bones; one ischio-pubic bone and one femur. The humeri both bear the distal tuberosity referred to on p. 53, on their dorsal face; their length is 14·6 cm., width of distal end 13·2 cm. The ischio-pubic bone is much shorter and broader than usual, and the upper end is more enlarged; the length is 11 cm., width of upper end 6·4 cm., width of lower end 7·5 cm. The femur has a greatly enlarged bulbous proximal end, apparently owing to the development of exostoses: its length is 9·1 cm., width of distal end 5·9 cm.

R. 2162 (Leeds Coll. 31). Bones of back of skull, including basioccipital (text-fig. 1), basisphenoid (text-figs. 1 & 5), supraoccipital (text-fig. 2, C, D), right exoccipital (text-fig. 2, A, B), right and left stapes (right, text-fig. 3, C, D), right and left opisthotic (right, text-fig. 3, E, F), one pro-otic, left quadrate, right parietal (text-fig. 14).

The dimensions (in centimetres) of some of these bones are:—

Basioccipital: width of condyle	6·8
total length	7·2
„ width	8·8
length of neural surface	4·5
Exoccipital: height	3·5
length of base	4·7

70   MARINE REPTILES OF THE OXFORD CLAY.

> Supraoccipital : extreme width . . . . 7·5
> „ height . . . . 4·9
> width of neural opening at lower end . . . 1·2
> width between inner edges of the foramina
> marked (*for.*) in text-figure 2, C, D . . . 5·9
> Stapes : greatest length . . . . 7·6
> „ thickness of inner end . . . . 5·2
> Opisthotic : greatest length . . . . 7·8
> „ width of inner end . . . . 4·7
> Quadrate : greatest length . . . . 12·3
> „ width of articular surface . . (approx.) 4·5
> Basisphenoid : greatest length . . . . 7·5
> „ width . . . . 11·0
> Parietal : total length in straight line . . . . 17·2
> greatest width . . . . 6·5

R. 2161 (Leeds Coll. 90). Bones of the back of the skull, including the basioccipital, exoccipital, supraoccipital, stapes, imperfect opisthotics, one pro-otic (text-fig. 3), basisphenoid (text-fig. 6, B).

> The dimensions (in centimetres) of some of these bones are :—
>
> Basioccipital : width of condyle . . . . 6·7
> total length . . . . 6·5
> „ width . . . . 9·0
> length of neural surface . . . . 4·3
> Exoccipital : height . . . . 3·3
> Supraoccipital : extreme width . . . . (approx.) 8·0
> height . . . . 4·9
> width of neural opening at lower end . . . ·9
> width between inner edges of the foramina
> marked (*for.*) in text-fig. 2, C, D . . . . 5·0
> Stapes : greatest length . . . . 7·4
> „ thickness of inner end . . . . 5·1
> Basisphenoid : greatest length . . . . 6·4
> „ width . . . . 9·2

R. 3013 (Leeds Coll. 91). A somewhat crushed skull and mandible. The bones of the occipital region and palate are concealed or displaced, but the roof is fairly well preserved : a few small teeth (Pl. I. figs. 1–6) and portions of sclerotic plates are preserved. Also seven caudal vertebræ, a neural arch, numerous fragments of ribs, the right coracoid, and scapula. All the bones are fully ossified and seem to have belonged to an old individual.

> The dimensions (in centimetres) are :—
>
> Skull :
> Approximate length . . . . 102·0
> Length from anterior angle of external nares to tip of snout 50·0
> Length of temporal fossa . . . . 13·0
> Width of temporal fossa . . . . (approx.) 7·5
> Width between posterior angles of squamosal . (approx.) 28·0

OPHTHALMOSAURUS ICENICUS.

Tooth: length of crown	1·5
diameter at base of crown	·6
Coracoid: length	22·6
width	19·3
Scapula: length	24·0
width at narrowest part of blade	4·6
„ of proximal expansion	13·8

R. 2185. Portions of the skull of an old individual. The bones preserved are the quadrates with the adherent portions of the pterygoid and, on one side, of the stapes, the left opisthotic, one pro-otic, portions of the squamosal, and the parietals. With these are associated parts of the mandible, the centra of three precaudal and three caudal vertebræ, fragments of the coracoids, the right humerus, both radii, one ulna, and thirty-six other paddle-bones, also one femur. Most of the bones are crushed and distorted.

The dimensions (in centimetres) of some of the bones are :—

Quadrate: length	11·2
width at neck	6·3
Stapes: length	6·6
Opisthotic: length	6·9
Coracoid: width (approx.)	19·5
Humerus: length	17·5
Femur: length	11·1

R. 2191. Portions of skull and mandible, hyoid, and some fragments of ribs. Of the skull the nasals, one maxilla, one lachrymal are the best-preserved elements; in the mandible one of the articulars, the splenials, and some other parts are present. The length of the hyoid is 16·7.

R. 2155. Portions of skull and mandible, including the basioccipital, basisphenoid, both stapes, opisthotic, both articulars, and fragments. Also the fused atlas and axis, with twenty other precaudal centra and right humerus.

The dimensions (in centimetres) of some of these bones are:—

Basioccipital: greatest length	7·1
„ width	8·5
width of occipital condyle	6·8
Basisphenoid: greatest length	7·6
„ width	9·8
Stapes: greatest length	7·5
width of inner end	5·6
Atlas and axis: length	4·5
width	8·6
height	8·2
Humerus: length (approx.)	16·5

R. 2132 (Leeds Coll. 61). Includes the quadrates (imperfect), some fragments of skull-bones and sclerotic plates, twenty-six precaudal and thirty-one caudal vertebral centra, all well preserved and undistorted, the coracoids, imperfect scapulæ, humeri, radius, intermedium, pisiform, and twenty-three other paddle-bones, a femur, and numerous fragments of ribs.

The length of the humerus is 15 cm., that of the femur 11 cm. The width of the narrowest part of the shaft in the former is 7 cm., in the latter 3·9 cm.

R. 2150 (Leeds Coll. 77). Parts of skull, including basioccipital and imperfect quadrates, numerous loose teeth, seventy-five greatly-crushed vertebræ, three caudal neural arches, imperfect coracoid, scapulæ, humerus, and twenty-nine paddle-bones.

The length of the humerus is 14·4 cm., the width of its distal end 12·2 cm. The length of the scapula is 18·2 cm., the width of its distal expansion 10·8 cm. Some of the teeth are 1·9 cm. long, with a maximum diameter of 1 cm. at about the middle of the root.

R. 2153. Fused atlas and axis, twelve other precaudal and three caudal vertebræ, eight odd paddle-bones.

The dimensions (in centimetres) of some of the bones are:—

Atlas and axis : length of centra	3·4
depth of centra	6·1
width of centra	6·6
Largest caudal vertebra : length of centrum	3·7
depth of centrum	10·2
width of centrum	9·5

R. 2150 a. Atlas and axis, figured in text-fig. 24, A, B. The length of the combined centra is 4·5 cm., the greatest height 8·6 cm., the greatest width 8 cm.

R. 2152. A basioccipital, atlas and axis (text-fig. 24, E, F), twenty-one other precaudal and fifteen caudal centra. The caudal centra have been strongly compressed longitudinally, so that they are not much more than half their original length, but are not otherwise distorted. Also the coracoids, left scapula, part of left humerus, and six paddle-bones.

The dimensions (in centimetres) of some of these bones are:—

Basioccipital : greatest length (approx.)	6·6
„ width	9·0
width of condyle	6·1
Atlas and axis : length	3·6
width	8·6
height	8·3
Coracoid : width (approx.)	21·0
Scapula : length	21·3
width of lower expansion	14·6

R. 2143 (Leeds Coll. 82). The left quadrate, right scapula, the centra of three posterior precaudal and twenty-two anterior caudal vertebræ of a large individual. The centrum of vertebra which appears to be the first caudal is 10·7 cm. high, 10·7 cm. wide, 4·2 cm.

OPHTHALMOSAURUS ICENICUS.

long, approximately. The quadrate is massive and very extensively ossified; its extreme length is 13·5 cm., the width of the neck 6·8 cm.

The scapula is very fully ossified, and the deflected antero-inferior angle forms a definite and prominent process (? acromium). The length is 23 cm., the width of the proximal expansion 14·8 cm.

R. 2173 (Leeds Coll. 56). A caudal vertebra, a left humerus with radius, ulna, and seventeen other paddle-bones. The humerus has the prominences on the anterior angles of its distal end strongly developed on both its upper and lower surfaces; the pisiform facet is very small; its length is 18·5 cm., the width of its distal end 15·4 cm.

R. 2174. Atlas and axis with twenty-seven other precaudal vertebræ and a femur of a young individual. The original line of separation between the fused atlas and axis centra is clearly marked, even on the floor of the neural canal, where in most cases it is early obliterated.

The dimensions of the combined atlas and axis centra are: height 5·3 cm., width 6·0 cm., length at neural canal 2·7 cm. The femur is only 4·7 cm. long, its distal end being 3·2 cm. wide.

R. 2163. Bones of the back of the skull of a small individual, including the basioccipital, basisphenoid, stapes, one opisthotic, and quadrates.

R. 2135 (Leeds Coll. 71). A series of vertebræ, including the atlas, axis, and thirty other precaudal centra, together with twenty-four caudals from various parts of the tail. Also portions of the coracoids, the right humerus, both radii, an intermedium, a pisiform, and fifteen other paddle-bones. Some of these last are very thin and seem to have belonged to the edge of the paddle; their surfaces also are broken up by grooves into a number of irregular areas with smooth surfaces, which may indicate that they were merely covered with some sort of horny epidermal structure. The atlas and axis, with the four succeeding vertebræ, are shown in text-fig. 25.

Some dimensions (in centimetres) of these specimens are:—

Atlas and axis: length of centra	4·8
width of centra	8·1
depth of centra	8·6
Third cervical: length	2·8
width	6·6
depth	7·4
Fourth cervical: length	2·8
width	6·3
depth	6·9
Total length of the six anterior vertebræ (text-fig. 25)	17·0
Humerus: length	19·7

R. 2188. Portion of basioccipital. About eighty-six vertebral centra, mostly crushed and broken, with the exception of the posterior caudals, about fifty-four in number. Also a femur and sixteen paddle-bones. The chief peculiarity about this skeleton is, that the basioccipital seems to have been formed by two centra, an anterior and a posterior, the posterior being procœlous, the convex hinder surface forming the occipital condyle, the anterior concave face closely resembling the face of a vertebral centrum of ordinary type.

R. 2148 (Leeds Coll. 69). Fragments of mandible, three anterior cervical centra, and fifteen caudal centra, including some from the bend of the tail and some of the small terminal bones, the coracoids, and a femur of a young individual.

The dimensions (in centimetres) of some of the bones are :—

Cervical vertebral centrum : length	2·0
height	4·4
width	5·0
Caudal centrum from bend of tail : length (at top)	1·5
,, (at bottom)	1·0
height	3·0
width	2·6
Coracoid : length	18·0
breadth	16·0
Femur : length	7·6

R. 2137 (Leeds Coll. 63). Eleven posterior precaudal vertebral centra, three neural arches, many ribs, the complete shoulder-girdle (coracoids, scapulæ, clavicles, interclavicle), one complete ischio-pubic bone and the proximal end of the other. All these bones, which are those of a large individual, are uncrushed and exceptionally well ossified. The shoulder-girdle has been figured by Prof. H. G. Seeley in Proc. Roy. Soc. vol. liv. (1893) p. 151, fig. 1 ; see also text-figs. 32–35.

The dimensions (in centimetres) of some of the bones are :—

Dorsal vertebral centrum : length	4·5
height	8·4
width	10·2
Neural arch : total height	11·6
width of spine	3·9
Coracoid (text-fig. 32) : greatest length	23·0
,, width	20·3
length of symphysial surface . (approx.)	14·5
depth of symphysial surface . (approx.)	7·1
length of glenoid surface . . (approx.)	10·0
Scapula (text-fig. 33, A & B) : extreme length	24·0
width of blade at narrowest	4·4
,, expanded lower end	16·0
length of glenoid surface	6·0
width of glenoid surface	4·0
Clavicle (text-fig. 34) : length along outside of curve	37·5
,, in straight line	34·5
greatest width (in middle)	5·7

Interclavicle (text-fig. 34): greatest length in middle line . . 13·8
length of cross-bar . . (approx.) 23·0
Ischio-pubis: length . . . . 14·7
width of proximal end . . . . 5·1
„ distal end . . . . 7·1

R. 2147 (Leeds Coll. 68). Centra of fifteen precaudal vertebræ, nine neural arches, and parts of the clavicular arch including the median portions of both clavicles, the right one closely united with the corresponding portion of the interclavicle. Both clavicles seem to be somewhat deformed; that on the right side may have been fractured and afterwards mended during the animal's life.

R. 2141. Twenty-two precaudal, twenty-nine anterior caudal, and eight posterior caudal centra of a small individual. Some of the posterior caudals are from near the end of the tail and are very small. The dimensions of one are: length 1 cm., width 1·25 cm., height 1·3 cm.

R. 2157 (Leeds Coll. 29). Centra of twelve precaudal and nineteen caudal vertebræ of an old individual.

R. 2139 (Leeds Coll. 73). Centra of eleven vertebræ, six neural arches, one scapula, one coracoid.

R. 2164. Basioccipital, basisphenoid (text-fig. 6, A), stapes. The parasphenoid extends round the sides of the carotid foramen.

The dimensions (in centimetres) of these bones are:—

Basioccipital: greatest length . . . . 6·4
„ width . . . . 8·7
Basisphenoid: greatest length . . . . 7·0
„ breadth . . . . 9·0
Stapes: length . . . . 7·2
width of inner end . . . . 4·9

R 2134. Right fore paddle. Co-type described and figured by Seeley in Quart. Journ. Geol. Soc. vol. xxx. (1874) p. 703, pl. xlvi. fig. 3.

The dimensions (in centimetres) of this specimen are:—

Humerus: length . . . . 17·6
width at proximal end . . . . 12·7
„ middle of shaft . . . . 8·0
„ distal end . . . . 14·3
Radius: length . . . . 7·6
width . . . . 7·0
Ulna: length . . . . 5·9
width . . . . 7·6

76   MARINE REPTILES OF THE OXFORD CLAY.

R. 2175. A series of vertebral centra and a right coracoid of a young individual. The vertebræ include the atlas and axis (text-fig. 24, C & D), the line of junction of which is shown by a sharply-defined groove which is only interrupted at the floor of the neural canal; the axis has distinct diapophysial and parapophysial facets; the position of the intervertebral wedge-bones is shown by the facets with which they united. There are twenty-nine other precaudal vertebræ and twenty-eight caudals. The small coracoid presents no remarkable characters.

The dimensions (in centimetres) of some of the bones are :—

Atlas and axis: combined length	2·6
depth of centra	4·2
width of centra	4·9
Anterior caudal vertebra: length of centrum	2·2
depth of centrum	5·1
width of centrum	5·3
Coracoid: greatest width	9·9
,, length	11·8

R. 2169 (? Leeds Coll. 80). Twenty-four caudal centra; of these, one from the bend of the tail and another further back are figured (text-fig. 28). Also three neural arches.

The dimensions (in centimetres) of the first figured centrum and arch are :—

Centrum: length at neural canal	2·1
,, ventral edge	1·7
width	3·7
height	4·1

R. 3533. Portions of a young skull and skeleton, including basioccipital, exoccipital, supraoccipital, stapes, opisthotics, quadrates, pterygoids, squamosals, maxillæ, nasals, vomers (text-fig. 19), and other bones of the skull and mandible. Sixty vertebral centra, some neural arches and ribs.

R. 3535. Bones of skull and skeleton of a large individual, including parietals, frontals, and squamosals united (text-fig. 15), basisphenoid with parasphenoid, quadrates, stapes, opisthotic, portions of pterygoid, two vertebral centra; also the clavicular arch. Numerous sclerotic plates.

R. 3534. Eighteen centra of caudal vertebræ, mostly with the neural arches and ribs associated with them (text-fig. 27). Also imperfect shoulder-girdle, fore paddles, imperfect pelvis, hind paddles.

NOTE.—In the explanation of fig. 40 (p. 59), *for* "Left ischio-pubis of *Ophthalmosaurus*: A, from outer side; B, proximal end; C, from inner side," *read* "Right ischio-pubis of *Ophthalmosaurus*: A, from inner side; B, proximal end; C, from outer side."

Order SAUROPTERYGIA.

Carnivorous aquatic reptiles in which the skull has only one temporal arcade and a fixed quadrate; the pterygoids extending forwards to meet the vomers (? in all); the external nares situated some distance behind the end of the snout; a pineal foramen present. The teeth are thecodont, and sometimes a few are considerably enlarged; in the later forms they are confined to the edges of the jaws. The tail is relatively short, swimming having been effected mainly by the limbs, which become paddle-like; the hind pair never greatly reduced. Dorsal ribs with a single head; a plastron of ventral ribs.

Suborder PLESIOSAURIA.

Clavicular arch undergoing reduction and tending to become situated on the visceral side of a ventral extension of the scapulæ, which in the later types replace it functionally; coracoids large. The ilium is directed backwards and articulates with the ischium only; the pubis is a broad plate of bone, and the ischia also are generally much expanded. The limbs form oar-like paddles. The plastron consists of a median and several lateral series of overlapping ventral ribs.

Family ELASMOSAURIDÆ.

Head relatively small; neck long, in some cases excessively so. Cervical ribs with single head. Scapulæ meeting in the middle line, where they join the corresponding median anterior prolongations of the coracoids, at least in fully adult individuals. Clavicles and interclavicles may both be present, but one or both are usually greatly reduced. Epipodial bones much modified, being shortened up so as to resemble mesopodials.

Middle Jurassic to Cretaceous of Europe, North America, and perhaps New Zealand.

Genus **MURÆNOSAURUS**, Seeley.
[Quart. Journ. Geol. Soc. vol. xxx. (1874) p. 197.]

Skull short and broad, of relatively small size. About 24 teeth on each side in the upper jaw, five being situated in the premaxilla; of the maxillary teeth, the third, fourth, and fifth are enlarged. Mandible with a short symphysis and bearing about 20 teeth on each side. Neck consisting of about 44 vertebræ, the centra of which in the anterior part are about as long as broad. In the shoulder-girdle there is a well-developed interclavicle, while the clavicles are generally greatly reduced, in some cases being mere films of bone adherent to the visceral face of the interclavicle; in some cases probably they are wanting entirely. Coracoids not greatly produced outwards and backwards into postero-lateral processes. Fore limb a little larger than the hind limb, to which it is very similar in form, the humerus not being greatly expanded at its distal end even in the adult.

Middle Jurassic.

The specimen upon which Prof. H. G. Seeley founded *Murænosaurus leedsi*, the type species of the genus, is included in the Leeds Collection (No. 25, R. 2421); it consists of portions of the skull and mandible, 79 vertebræ, some of the caudals having been lost, numerous ribs, coracoids, scapulæ, pubes, ischia, ilia, and both the fore and hind paddles. The following account of the skeleton in this genus is founded so far as possible on this specimen, but many other nearly complete skeletons of the same or closely similar species have been employed to supplement the description.

Skull (Pl. III.; Pl. VI. figs. 1-2; text-figs. 43-47).—The skull is small in proportion to the size of the animal, and is roughly triangular in outline, the muzzle being bluntly pointed. The upper surface of the anterior portion was probably gently convex from side to side, while in the parietal region there is a high, sharp, sagittal crest, the posterior end of which is the highest point of the skull, from which it slopes gradually down to the tip of the snout. The following account of the individual bones is founded on several more or less nearly complete examples—the best (R. 2678) belonging to the skeleton which is the type specimen of *M. platyclis*. Portions of the skull of *M. durobrivensis* (R. 2861), as well as of the type of *M. leedsi*, are also figured and described.

The *basioccipital* (*b.oc.*, Pl. III. figs. 1, 1 *a*; text-figs. 43, 44) bears the whole of the nearly hemispherical occipital condyle (*oc.c.*), the border of which forms a sharp rim, sometimes separated by a short neck from another parallel rim (text-fig. 44); the upper border of the condyle in some specimens is a little flattened beneath the neural canal; there is no pit marking the original position of the notochord, such as has been described as occurring on the occipital condyle of *Ophthalmosaurus* (see *supra*, p. 6). The upper surface of the bone (text-fig. 43, B) bears a pair of large elongated oval facets (*exo.f.*), for union with the ventral end of the exoccipitals and opisthotics; the space between these facets (*n.c.*), forming the floor of the brain-case, is narrow and concave from side to side behind, while in front it widens out and bears in its middle line a strong longitudinal ridge which in some specimens is paired. The anterior face of the bone (text-fig. 43, C) is nearly vertically truncated by the surface (*bs.f.*) for union with the basisphenoid. Anterolaterally the body of the bone bears a pair of stout processes which project outwards and downwards, their ends being truncated by large oval facets (*pt.f.*) which look outwards and forwards and articulate with the pterygoids, which in these animals extend very far back, their posterior ends, by which they unite with the quadrates, being actually behind the occipital condyle. Between these processes the ventral surface of the bone is concave from side to side; in front of this it is flattened and greatly roughened up to its anterior edge. In some cases the posterior surface of the lateral processes bears a roughened surface for the attachment of muscles.

Text-fig. 43.

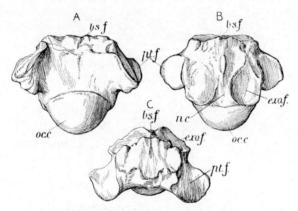

Basioccipital of *Muraenosaurus durobrivensis*: A, from below; B, from above; C, from front. (R. 2861, nat. size.)

bs.f., facet for basisphenoid; *ex.of.*, facet for exoccipital; *n.c.*, floor of neural canal; *oc.c.*, occipital condyle; *pt.f.*, facet for pterygoid.

The *basisphenoid* (*b.s.*, Pl. III. figs. 1, 1 *a*; text-fig. 44) consists of a posterior cuboid mass of bone, the antero-ventral portion of which is prolonged forwards to form the floor of the pituitary fossa. The posterior face is nearly flat and fitted closely against the corresponding surface of the basioccipital. In most cases the two bones can be seen to make a very obtuse angle with one another, the axis of the basis cranii in front of the basioccipital being bent a little upwards. The nearly flat upper surface consists of two flattened areas united in front, but separated posteriorly by a **V**-shaped concavity; in life these areas were perhaps covered with

80 MARINE REPTILES OF THE OXFORD CLAY.

cartilage. In front of this region is the deeply concave pituitary fossa. The sides of the bone are flat and roughened posteriorly, and bear deep grooves running downwards and forwards from the upper posterior angles and terminating in the large paired foramina (*i.c.f.*), by which the internal carotids enter the pituitary fossa. Above these foramina the sides of the bone are produced outwards and forwards

Text-fig. 44.

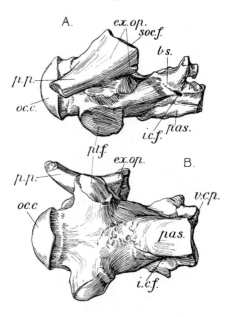

Basioccipital, exoccipital-opisthotic, basisphenoid, and part of parasphenoid of *Murænosaurus lædsi*: A, from side; B, from below. (R. 2422, nat. size.)

bs., basisphenoid; *ex.op.*, united exoccipital and opisthotic; *i.c.f.*, internal carotid foramen; *oc.c.*, occipital condyle; *pas.*, parasphenoid; *p.p.*, paroccipital process; *pt.f.*, pterygoid facet; *soc.f.*, facet for supraoccipital; *v.c.p.*, lower cylindrical processes of basisphenoid.

into a pair of processes bearing at their extremities facets, presumably for union with the pterygoids. The ventral surface is much roughened posteriorly, but for the greater part of its extent it is concealed by the closely adherent hinder end of the parasphenoid (*pas.*). The posterior border of this is sometimes notched in the middle line (Pl. III. fig. 1) or may be squarely truncated; in the former case the appearance of the presence of a median foramen is produced, and there may indeed be a small nutritive foramen at this point, but, as already described, the paired internal carotids enter the pituitary fossa by lateral foramina, not by a median ventral opening as in *Ophthalmosaurus* (see p. 13, fig. 5). The anterior end of the basisphenoid beneath and in front of the pituitary fossa terminates in a pair of blunt truncated processes (*v.c.p.*, the lower cylindrical processes of Siebenrock); these probably joined the cartilaginous presphenoid region.

The *parasphenoid* (*pas.*, Pl. III. figs. 1, 1 *a*; text-fig. 44) is a thin flat bone, the hinder part of which is closely united with the ventral face of the basisphenoid, though the line of junction usually remains distinctly visible. In front of this it divides the two openings which in former papers * I have called the posterior palatine vacuities (parasphenoidal vacuities of Williston), which are enclosed externally by the pterygoids. Anteriorly the parasphenoid widens out somewhat, and, judging from what occurs in the closely allied genus *Tricleidus*, the outer edges were overlapped for some distance by the pterygoids. In front of this union the bone is abruptly truncated, its border forming the hinder limit of the median interpterygoid vacuity. This form of parasphenoid differs widely from that described by Williston† in *Trinacromerum*, where it is compressed laterally, and is so deep posteriorly that its upper surface would seem to be on a level with the cranial surface of the basisphenoid and basioccipital, there apparently being no pituitary fossa. Moreover, according to Williston, the pterygoids meet in median suture above the hinder end of the parasphenoid, so that its posterior end unites with them and not with the ventral face of the basisphenoid as usual.

The *exoccipital* (*exo.*, Pl. III. fig. 1 *a* ; text-figs. 44 & 45, C, D, E) in all cases observed is fused with the opisthotic, though on the inner face of some specimens the line of junction of the two elements is quite distinct, as it is also in some instances on the articular ends (see dotted line in text-fig. 45, E). The bone formed by the united elements is columnar in form, and the exoccipital portion forms the sides of the foramen magnum. The lower end expands somewhat and terminates in an oval, flattened, more or less roughened facet for union with the basioccipital (*boc.f.*). The anterior angle of this surface is borne by the lower end of the opisthotic, which is here marked off from the exoccipital in many cases by a groove (*op.g.*), and on the inner side by a distinct notch. The inner face of the combined bones is concave from above downwards, and is perforated by a number of openings. Of these the

---

* See figures of skulls of *Plesiosaurus* in Quart. Journ. Geol. Soc. vol. lii. (1896) p. 251, fig. 2, and pl. ix. fig. 1 ; of *Pliosaurus, loc. cit.* vol. liii. (1897) pl. xii. ; of *Peloneustes* in Ann. Mag. Nat. Hist. [6] vol. xvi. (1895) p. 245, and pl. xiii. fig. 1.

† Williston, "North American Plesiosaurs: *Trinacromerum*," Journal of Geology, vol. xvi. (1908) p. 718, fig. 5.

two posterior (XII.) probably transmitted branches of the hypoglossal nerve; the hindermost is the larger and its outer opening is at the posterior end of the deep groove beneath the base of the paroccipital process (*p.p.*). In front of these foramina there is on the inner face of the bones a large funnel-shaped aperture (*jug.*), opening externally at the anterior end of the above-mentioned groove. This aperture corresponds with the so-called jugal foramen of Siebenrock * and transmitted the vagus group of nerves. It lies between the exoccipital and the opisthotic elements, the latter in this case sending down a process which fuses with the exoccipital beneath the foramen and meets the basioccipital as above described. According to Siebenrock † the opisthotic has a similar articulation with the basioccipital in *Hatteria*, but this is denied by Osawa ‡.

Above and a little in front of the jugular foramen, and separated from it by a sharp ridge forming the hinder edge of the opisthotic, there is a deep rounded fossa (*a.*) for the ampulla of the posterior vertical semicircular canal (*p.v.c.*), the channel for which passes outwards and upwards from the inner end of the fossa, opening on the upper surface by which the bone unites with the supraoccipital (*soc.f.*). Another passage also runs outwards and forwards from the fossa and opens on the surface for union with the pro-otic; this is the channel for the horizontal semicircular canal (*h.c.*). The upper end of the combined exoccipital and opisthotic bears a large flattened ovate surface for union with the supraoccipital (*soc.f.*); this facet, which looks directly upwards, is mostly borne by the opisthotic and, as mentioned above, is perforated for the passage of the posterior vertical semicircular canal (*p.v.c.*). Anteriorly the opisthotic bears another facet (*pro.f.*) looking upwards and forwards, making an angle of about 60° with the supraoccipital surface; this facet is for union with the pro-otic, and is perforated by the channel of the horizontal semicircular canal (*h.c.*).

The outer face of the combined bones is somewhat flattened superiorly, while its lower portion bears the greatly elongated paroccipital, the base of which extends across the anterior two-thirds of the bone, being probably entirely borne by the opisthotic. The process itself is directed downwards and backwards; it narrows somewhat in the middle, but expands distally and terminates in a roughened convex facet (*q.f.*), by which no doubt it articulates with the quadrate, although no specimen showing the actual junction has been observed. Beneath the base of the paroccipital process is a deep groove, at the posterior end of which is the outer opening of the passage for the XII. nerve, while at the anterior end is the jugular

---

* Siebenrock, "Zur Anatomie des *Hatteria*-Kopfes," Sitz. Akad. Wissensch. Wien, vol. cii. pt. i. (1893) p. 256.

† *Op. cit.* p. 250.

‡ Osawa, "Beiträge zur Anatomie des *Hatteria punctata*," Arch. f. Mikroskop. Anatomie, vol. li. (1897–8) p. 495.

foramen (*jug.*). The posterior (exoccipital) edge of the combined bones is sharp; the anterior (opisthotic) broad, rounded, and concave from above downwards, the upper end of the concavity being continued as a shallow groove (*c.a.*) on to the base of the paroccipital process; this groove no doubt marks the position of the columella auris. On the inner face of the bone there is a small foramen (*f.*), probably for a blood-vessel.

The *supraoccipital* (text-fig. 45, A, B) is a short ∩-shaped bone which enclosed the

Text-fig. 45.

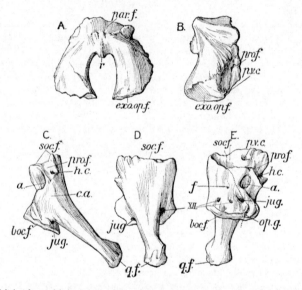

Supraoccipital and exoccipital-opisthotic of *Muraenosaurus durobrivensis*: A, supraoccipital from behind; B, from right side; C, exoccipital-opisthotic from front; D, exoccipital-opisthotic from outer side; E. exoccipital-opisthotic from inner (cranial) side. (R. 2861, nat. size).

a., cavity for ampulla of posterior vertical semicircular canal; *boc.f.*, facet for basioccipital; *c.a.*, groove for columella auris; *exo.op.f.*, facet for exoccipital-opisthotic; *f.*, foramen (? vascular) on inner face of exoccipital; *h.c.*, channel for horizontal semicircular canal; *jug.*, jugular foramen; *op.g.*, groove marking line of junction between the exoccipital and opisthotic elements; *par.f.*, facet for union with the parietal; *pro.f.*, facet for pro-otic; *p.v.c.*, channel for posterior vertical semicircular canal; *q.f.*, facet for quadrate; *r.*, ridge projecting into upper end of the foramen magnum; *soc.f.*, facet for supraoccipital; XII., foramina for the hypoglossal nerve.

upper half of the foramen magnum. From the upper border of the arch a sharp ridge-like process (*r.*) projects downwards into the foramen. The ventral ends of the ∩ are greatly expanded and terminate in a roughly triangular surface looking downwards, for union with the exoccipital and opisthotic (*exo.op.f.*), and a large surface looking forwards and downwards for union with the pro-otic (*pro.f.*); this latter facet bearing near its inner border two foramina united by a nearly closed channel which probably lodged part of the posterior vertical semicircular canal (*p.v.c.*). The summit of the bone is occupied by a flattened and roughened surface, for union with the united parietals (*par.f.*), on the ventral face of which there is a corresponding slightly concave facet.

No good specimen of the *pro-otic* has been found, but in one instance (No. 18, R. 2861) it is preserved on one side, still united with the supraoccipital, but crushed inwards. It appears to have been roughly triangular in outline, each of the angles being truncated. Its upper anterior edge is broadly rounded; behind this it unites by a large surface with the supraoccipital and below this with the opisthotic, forming with these bones the usual triradiate suture; at its lower anterior end there is a large articular facet, perhaps for union with the basisphenoid. The inner face of the bone is hollowed into a deep fossa, from which two channels diverge about at right angles; these probably lodged portions of the anterior vertical and the horizontal semicircular canals.

The *parietals* (*par.*, Pl. III. fig. 2 *a*; Pl. VI. fig. 1; text-fig. 46) form the whole of the cranial roof behind the parietal foramen; they seem to fuse completely with one another, though traces of their original separation can be seen, particularly on the occipital surface, where a slight vertical ridge marks the line of union. They unite with the supraoccipital by a rounded and slightly concave surface, and above this they rise for some distance, forming a nearly vertical occipital surface, concave from side to side. In this region the bones are produced laterally into short stout triangular squamosal processes, which are completely overlapped along their upper borders by the parietal processes of the squamosals, which meet in the middle line and unite closely with one another, forming the actual vertex of the occipital surface of the skull. In front of the squamosal processes the upper part of the parietals narrows into a high sagittal crest, which extends forwards to within two or three centimetres of the pineal foramen (*p.f.*). Here the bones widen out considerably and enclose the large foramen, uniting immediately in front of it with the frontals by a complex suture running nearly transversely. In some cases this suture seems to cross the front of the pineal foramen, in which case the frontals take a small share in the formation of the anterior border of that opening. Externally the anterior expansion of the parietal joins the postfrontal (*po.f.*) in a nearly longitudinal suture. Beneath the sagittal crest the parietals widen out to form the cranial roof, their greatest width being about opposite the hinder end of the pineal foramen, where on the ventral surface in one specimen there are traces of union with the upper end of the columella. For about 2 centimetres in front of the surface for union with the supraoccipital the cranial surface is concave in all directions; anteriorly it passes into the broad funnel-shaped inner opening of the pineal foramen, the sides of which are formed by strong ridges which are continued forwards on to the ventral face of the frontals (Pl. III. fig. 2).

Text-fig. 46.

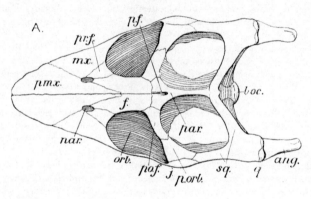

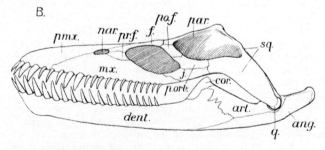

Semi-diagrammatic restoration of the skull and mandible of *Murænosaurus*: A, from above; B, from left side. (About ¼ nat. size.)

ang., angular; *art.*, articular; *boc.*, basioccipital; *cor.*, coronoid process of splenial; *dent.*, dentary; *f.*, frontals; *j.*, jugal; *mx.*, maxilla; *nar.*, external nares; *orb.*, orbit; *par.*, parietals; *p.f.*, pineal foramen; *pmx.*, premaxillæ; *po.f.*, postfrontal; *p.orb.*, postorbital; *pr.f.*, prefrontal; *q.*, quadrate; *sq.*, squamosal.

The *frontals* (*f.*, Pl. III. figs. 2, 2 *a*; Pl. VI. fig. 1; text-fig. 46) are large, and there may be some doubt whether the bones so called here may not include other elements, though in no case can any trace of sutures be observed. Posteriorly they join the parietals in a transverse suture and, as above mentioned, may or may not form the anterior edge of the pineal foramen. External to their union with the parietals they join the postfrontals for a short distance. In front of this, again, they widen out by a gentle curve, forming the roof of the orbit. In this region the upper surface of the two bones is convex from side to side laterally and concave mesially, where they unite with one another in the median line, so that here the roof of the skull is concave from side to side.

In front of the orbit the frontal attains its greatest width, sending downwards and outwards a process, which terminates in a sutural surface, presumably for union with the prefrontal, unless that element is actually included in the bone here called frontal. In front of the prefrontal process the bone again narrows and is notched by the inner border of the external narial opening. In the middle line as far back as the anterior third of the orbit the middle portion of the frontals is concealed beneath the long overlapping facial processes of the premaxillæ, which thus extend far back behind the external nares. On their ventral face (Pl. III. fig. 2) the frontals bear on either side a strong ridge or crest continuous posteriorly with the ridges on the parietals, referred to above. The ridges run parallel with one another and separated by only a narrow interval as far as opposite the anterior third of the orbital border, where they bifurcate, one branch running outwards on the preorbital region and helping to form the anterior wall of the orbit, the other continuing parallel with its fellow, on the lower face of that part of the frontal which is concealed by the facial processes of the premaxillæ. Between the divergent branches of the ridge there is a deep hollow, where the roofing-bone is very thin. In the orbital region the outer surface of the ridge is concave from below upwards as it passes to the orbital border, thus helping to form not only the roof but also the upper part of the inner wall of the eye-socket.

The *postfrontals* (*po.f.*, Pl. VI. fig. 1; text-fig. 46) stand out nearly at right angles to the long axis of the skull; at their inner end they unite posteriorly with the parietals, anteriorly with the frontals. The upper surface of these bones is concave posteriorly, but convex in front, the two areas being separated by a rounded ridge, the continuation outwards of the borders of the divisions of the sagittal crest, where they diverge on either side of the pineal foramen. On the outer end and on the outer portion of the posterior border are sutural surfaces, by which probably the bone joined the *postorbital* (*p.orb.*). The anterior edge of the bone constitutes the upper part of the posterior border of the orbit.

The *squamosal* (*sq.*, Pl. VI. fig. 1; text-figs. 46, 47), perhaps including the supratemporal, is badly preserved in all the skulls of *Murænosaurus* available for description. It is a triradiate bone; the upper, somewhat slender arm runs up to the lateral processes of the parietal, the upper edge of which it completely overlaps, meeting its fellow of the opposite side in the mid-dorsal line ; this bar, formed by the union of the parietal and squamosal, constitutes the hinder border of the temporal fossa. The lower arm is broad and unites closely with the quadrate. The anterior (zygomatic) arm, which is broad and flat, curves upwards and forwards, forming the posterior part of the single temporal arcade; in front it unites with the jugal and for a short distance along its upper border with a triangular element, the *postorbital (p.orb.)*. The ventral border of this bone joins the jugal, while its upper angle unites in suture with the postfrontal, together with which it forms the anterior boundary of the temporal fossa and the posterior border of the orbit. A ridge running near the posterior edge of the postorbital becomes continuous posteriorly with the upper edge of the zygomatic bar of the squamosal.

No well-preserved specimen of the *quadrate* (*q.*) region is available for description. The best shows that the articular surface for the mandible was strongly convex from before backwards and that from side to side it was concave on its outer, and strongly convex on its inner, half. There is no trace of a division into two elements, possibly quadrate and quadrato-jugal, both helping to form the articulation, such as will be noticed in the account of the skull of *Tricleidus*. The quadrate is firmly united on its inner anterior face with the posterior limb of the pterygoid, and externally is overlapped by the squamosal, but the precise manner in which these unions is effected cannot be seen.

The *jugal* (*j.*, Pl. VI. fig. 1; text-figs. 46, 47) is a short, broad, and roughly quadrangular bone ; posteriorly it unites with the zygomatic bar of the squamosal, above with the postorbital, below with the maxilla. The remaining anterior free border helps to form the rim of the orbit.

The *maxilla* (*mx.*, Pl. III. figs. 2, 2 *a* ; Pl. VI. fig. 1 ; text-figs. 46, 47) consists mainly of a strong alveolar region containing the sockets of 15–16 teeth, of which the third, fourth, and fifth are considerably the largest; behind these there is a diminution in size towards the hinder end of the series. On the inner side of each alveolus there is, as a rule, a pit, at the bottom of which the tip of the more or less developed successional tooth can be seen. The facial portion of the maxilla is a broad convex plate, forming the lower border of the orbit, to which, moreover, it seems to have supplied an imperfect floor. The palatal region of the bone is a thin plate uniting with the premaxillæ in front, then with the vomer, and behind this, again, with the palatines and probably also with the transverse bones. Posteriorly the maxilla joins the jugal, and in front of the orbital region its upper edge unites with the broad lower end of the prefrontal. The *prefrontal* (*pr.f.*) is not well preserved in any of the specimens. As just mentioned, its lower edge joins the maxilla, above which its posterior border helps to enclose the orbit. The relations of the anterior edges are obscure, but it appears that at its upper end it joins the frontal and anteriorly forms a small portion of the border of the external narial opening. If, as is possible, the true prefrontal is fused with the bones here described as frontals, then the element now under consideration must be called the lachrymal.

Each *premaxilla* (*pmx.*, Pl. III. figs. 2, 2 *a* ; Pl. VI. fig. 1 ; text-figs. 46, 47) consists of a broad anterior portion and a long facial process which extends back to behind the anterior border of the orbits. They unite in a close median suture ; in the broad anterior portion their upper surface is gently convex from side to side and is roughened and perforated with numerous vascular foramina, some near the middle line being of considerable size. The suture uniting these bones with the maxillæ crosses the alveolar border immediately behind the socket of the fifth tooth. On the facial surface the posterior border of the premaxilla runs upwards and backwards, passing into the outer border of the facial processes, a little below which it is notched

Text-fig. 47.

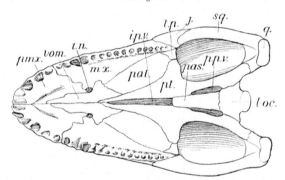

Semi-diagrammatic restoration of the palatal view of skull of *Murænosaurus*. (About ⅓ nat. size.)

boc., basioccipital ; *i.n.*, internal nares ; *i.p.v.*, interpterygoid vacuity ; *j.*, jugal ; *mx.*, maxilla ; *pal.*, palatine ; *pas.*, parasphenoid ; *pmx.*, premaxilla ; *p.p.v.*, posterior palatine vacuity ; *pt.*, pterygoid ; *q.*, quadrate ; *sq.*, squamosal ; *t.p.*, transpalatine ; *vom.*, vomer.

by the anterior ends of the external narial opening. On the palatal surface the anterior ends of the vomers are wedged in between the palatal plates of the premaxillæ, and extend as far forwards as the level of the alveolus of the third or fourth tooth. Each premaxilla carries sockets for five teeth : of these the first is small and close to the middle line ; the second, third, and fourth are much larger ; while the fifth is again small. The sockets of the first four teeth show that they were directed strongly forwards and doubtless were specially fitted for the prehension of living prey.

The *vomers* (*vom.*, text-fig. 47) occupy a considerable area in the anterior part of the palate. They consist of a comparatively narrow anterior portion, which is thrust between the premaxillæ, and a much broader posterior portion, the small internal nares being situated in the angle caused by the sudden widening. The outer and anterior border of the nares is formed by the maxillæ, which behind them unite with the vomers. Posteriorly the broad vomerine plates join the palatines and, towards the middle line, the pterygoids, the slender anterior ends of these bones being thrust between them for a short distance.

The *palatines* (*pal.*, text-fig. 47) are thin, roughly oblong bones; posteriorly they unite with the anterior edge of the lateral wing of the pterygoids (see below) and with the transpalatine, internally with the outer edge of the anterior limb of the pterygoids, while externally they join the maxillæ in an overlapping suture.

The *pterygoids* (*pt.*, text-fig. 47) are triradiate bones, of which the anterior and posterior rami are very long, the median one much shorter. The posterior ramus is stout and somewhat compressed laterally; towards its posterior end it becomes thicker, and on its inner face bears a large facet for union with the ventro-lateral process of the basioccipital described above. Behind this it curves outwards, and its hinder end unites with the inner edge of the quadrate, but no specimen showing the exact form of this part of the bone has been found. A little in front of the basioccipital facet there is another projection which seems to have effected a junction with the basisphenoid. The anterior (palatal) branch of the pterygoid is flattened and expanded. It is widest just in front of the point of union with the lateral ramus, and thence narrows gradually forwards terminating in a pointed process which unites with its fellow of the opposite side, the two running forwards between the hinder ends of the vomers. Behind this anterior union of the pterygoids there is in this genus a long median slit-like palatine vacuity extending back to the anterior end of the *parasphenoid* (*pas.*), which, judging from a well-preserved example in a closely allied genus, was short and abruptly truncated in front; the pterygoids seem to have overlapped the edge of the parasphenoid for a short distance. The outer edge of the anterior process of the pterygoid unites closely with the palatine. The lateral ramus is short; its posterior border is concave, passing by a gentle curve into the outer border of the posterior ramus. Towards its outer end it is much thickened to form a downwardly projecting knob, which unites with a corresponding process on the transverse bone. Anteriorly this ramus unites with the palatine. No good specimen of the transverse bone is preserved.

The structure of the *mandible* (Pl. III. figs. 3, 3 *a*; Pl. VI. fig. 2; text-figs. 46 & 48) is somewhat difficult to make out with certainty. This arises from the fact that some of the bones of the ordinary reptilian mandible are either wanting or, at least in all the specimens seen, fused with other elements. The largest bones of the mandible are, as usual, the *dentaries* (*dent.*); these form the whole, or nearly the whole, of the symphysis and the greater part of the rami to a little beyond the alveolar region. They are widest at the symphysis, where they form a slight expansion, bearing large teeth directed outwards and forwards; behind this the teeth become smaller and are directed more upwards. Throughout the alveolar region there is within the line of functional alveoli a series of pits, in which the tips of the replacing teeth appear. Within this again is a groove apparently running along the line of union of the dentary and splenial. There are about twenty teeth on each side of the mandible. The exact arrangement of the *splenials* (*spl.*) is not clear; they seem to form most of the inner surface of

Text-fig. 48.

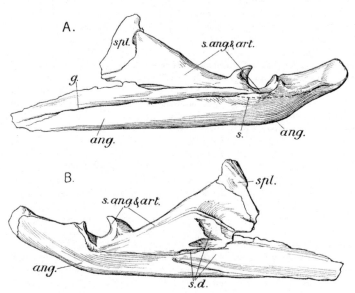

Posterior portion of right ramus of mandible of *Murænosaurus platyclis*: A, inner face; B, outer face. (R. 2678, type specimen, ⅔ nat. size.)

ang., angular; *g.*, groove on inner face of angular; *s.*, probable suture between angular and articular surangular; *s.ang. & art.*, united surangular and articular; *s.d.*, sutural surfaces for union with the dentary; *spl.*, coronoid process of splenial.

the mandibular ramus; their lower border is separated from the dentaries by a deep Meckelian groove, which is continued posteriorly on the inner face of the angular (see text-fig. 48, *g.*). They do not appear to have actually met in the symphysis (except perhaps for a short distance on the ventral surface), but are separated by a narrow wedge formed by the dentaries. Posteriorly it appears that they rise to form the anterior part and summit of the triangular coronoid process (text-fig. 48, *spl.*), and this region may include the fused coronoid bone, which has not been seen as a separate element in this genus, although it may be present in *Tricleidus*. The line of junction with the conjoined surangular and articular runs downwards and forwards from a point a little behind the summit of the coronoid process.

The posterior part of the mandible appears to be composed of only two separate pieces—one, the fused *articular* and *surangular* (Pl. VI. fig. 2; text-fig. 48, *s.ang. & art.*); the other, the *angular* (*ang.*). The combined surangular and articular form the upper and posterior part of the hinder end of the ramus; anteriorly the surangular unites in suture with the coronoid region of the splenial; below this on its outer face it bears deep depressions (text-fig. 48, *s.d.*) which receive the upper part of the posterior end of the dentary; below it unites with the angular in a long suture, which, as far as a point beneath the articular surface, runs parallel with the lower border of the jaw. At this point the suture turns up, and in the postarticular region runs along the upper and outer edge. The relations with the angular cannot be clearly seen on the inner face of the jaw, at least not posteriorly. The articular surface is very deeply concave from before backwards (more than a semicircle in *M. platyclis* figured in text-fig. 48); in its outer half it is bounded anteriorly by a very strong prominence projecting backwards and inwards. The surface, though concave from before back, is slightly convex from side to side, the median ridge thus formed working in the groove between the outer and inner condyles of the quadrate. No line of separation between the articular and surangular has been seen in any of the specimens examined. The angular (*ang.*) is a very large bone forming the whole of the lower part of the posterior half of the mandibular ramus. Its suture with the surangular and articular has been described above. In the postarticular region it forms a thin plate on the outer side and also the lower border and lower part of the inner face of the articular-surangular, but here the sutures are obscure. Possibly the inner half of the surface for the quadrate may be in part formed by the angular. In front of this surface the bone is deeply grooved longitudinally, the groove (text-fig. 48, *g.*) extending right forwards to its anterior extremity, where it receives the lower edge of the dentary, which overlaps it also on the outer side of the jaw, the surface of union being distinctly marked (text-fig. 48, *s.d.*).

The above description differs in several respects from that given by Williston * in the case of the mandibles of some American Plesiosaurs. In the mandible of *Polycotylus*, for instance, he recognises the presence of a suture between the articular and surangular, and also the occurrence of a preacticular element. He also, both in this genus and in the closely allied *Trinacromerum* †, finds a distinct coronoid element,

* "North American Plesiosaurs: *Trinacromerum*," Journal of Geology, vol. xvi. (1908) pp. 720–1, fig. 6.
† "North American Plesiosaurs, Pt. I.," Field Columbian Museum, Geological Series, vol. ii. no. 1 (1903) pp. 29–32, fig. 31.

of which nothing can be seen in *Murænosaurus*. Williston's * figure of the mandible of *Cimoliosaurus snowii* seems to show that in that species the structure was very similar to that above described.

Vertebral Column.—The vertebral column consists of 43–44 cervicals, 2 or 3 pectorals, 20 dorsals, 3–4 sacrals, and an unknown number of caudals. The cervical series forms the greater part of the column, and in the type specimen of *Murænosaurus leedsi* the neck was about 210 cm. long.

The *atlas* and *axis* (text-fig. 49) are always closely united in adult specimens, but in young individuals the lines of division between the different elements are quite distinct.

Text-fig. 49.

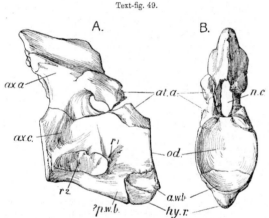

Atlas and axis of *Murænosaurus durobrivensis*: A, from right side; B, from front. (R. 2863, nat. size.)
at.a., neural arch of atlas; *a.w.b.*, anterior wedge-bone; *ax.a.*, neural arch of axis; *ax.c.*, centrum of axis; *hy.r.*, hypapophysial ridge; *n.c.*, neural canal; *od.*, odontoid; *?p.w.b.*, ? posterior wedge-bone; *r.¹*, rib of atlas; *r.²*, rib of axis.

The atlantal cup for the occipital condyle is deeply concave, and its diameter from above downwards is a little longer than from side to side. By far the greater part of it is formed by the anterior end of the odontoid (*od.*) process, which is here as large as or larger than the centrum of the axis. Its antero-ventral angle is cut away and unites with the anterior wedge-bone (*a.w.b.*), the front of which completes the lower portion of the atlantal cup. Above, the odontoid unites with the bases of lateral pieces of the neural arch of the atlas (*at.a.*). The union with these is not very extensive, since they do

* *Op. cit.* (1903) p. 52, fig. 13, and pl. v. fig. 5.

not meet in the middle line, the floor of the neural canal being formed by the odontoid ; and they only form a very small share of the supra-lateral borders of the atlantal cup, being widely separated from the ventral wedge-bone (text-fig. 49, B). In some Liassic Plesiosaurs, on the other hand, the odontoid takes a much smaller share in the formation of the atlantal cup, while the lateral pieces extend downwards and unite extensively with the upper end of the wedge-bone *, so that the odontoid only forms the middle portion and the neural border of the cup. Curiously enough, this condition is repeated in the American Cretaceous Plesiosaur *Trinacromerum* (*Dolichorhynchops*) *osborni*, described by Williston †. In fact, in the structure of the articulation with the skull, *Murænosaurus* seems to retain a much more primitive condition than is found in some, at least, of the Liassic and Cretaceous types.

The upper portions of the lateral (neural) pieces of the atlas run backwards and bear at their extremities comparatively large zygapophysial facets, by which they unite with the corresponding surfaces on the side of the arch of the axis (*ax.a.*). The two halves of the arch are separated from one another superiorly by a considerable interval. The anterior subvertebral wedge-bone (*a.w.b.*), as already noted, forms the lower portion of the atlantal cup. Above it unites with the odontoid in a suture running backwards and a little downwards. Posteriorly it is bounded by a nearly vertical suture, but whether this separates it from the second subvertebral wedge-bone (? *p.w.b.*) or from a ventral prolongation of the odontoid, cannot be determined ; its free ventral surface is raised into a strong hypapophysial ridge (*hy.r.*), which is continued back on the ventral surface of the bone behind, and dies away at the middle of the axial centrum (*ax.c.*). This latter is similar in every respect to the succeeding centra, except that it is fused with the odontoid in front. Above it bears a neural arch (*ax.a.*) with a spine, the highest point of which is at its posterior end. The pedicles of the arch widen out considerably at their ventral end, so that the surface for union with the centrum projects anteriorly a little beyond the axial centrum and unites also with the odontoid. The anterior and posterior zygapophyses are both well developed. Beneath the surface for union with the neural arch the sides of the centrum are a little concave, but then turn outwards and bear the facet ($r.^2$) for the large axial rib ; the extreme anterior edge of this facet seems to be borne by the odontoid, which also has a small prominence apparently indicating the presence of a very rudimentary atlantal rib ($r.^1$). The posterior face of the axial centrum is gently concave and is a rounded oval, the position of the neural canal being marked by a slight concavity.

The above description of the atlas and axis is founded on a specimen belonging to a

* See Barrett, "On the Atlas and Axis of a Plesiosaur," Ann. Mag. Nat. Hist. [3] vol. ii. (1858) p. 361, pl. xiii.

† "North American Plesiosaurs, Pt. I.," Field Columbian Museum, Geological Series, vol. ii. no. 1 (1903) p. 32, pl. xxii. fig. 5.

skeleton of *Murænosaurus durobrivensis* (R. 2863), since no well-preserved examples of these bones in the type species (*M. leedsi*) are contained in the collection. In *M. leedsi* the somewhat imperfect and badly-preserved atlas and axis have been described by Professor Seeley [*], and, so far as can be seen, differ in no important points from those just noticed (Pl. IV. fig. 1). The remainder of the cervical region (Pl. IV. figs. 2–4; Pl. V. figs. 1–3; Pl. VI. figs. 4, 5; text-figs. 50, 51) is characterised by the large number (about 41) of vertebræ composing it and the greater relative length of the centra compared with those of the cervicals in *Cryptocleidus* and

Text-fig. 50.

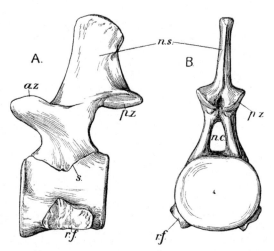

Anterior cervical vertebra of *Murænosaurus durobrivensis*: A, from left side; B, from behind. (R. 2863, about nat. size.)

a.z., anterior zygapophysis; *n.c.*, neural canal; *n.s.*, neural spine; *p.z.*, posterior zygapophysis; *r.f.*, facet for rib; *s.*, neuro-central suture.

Tricleidus: measurements will be given under the different species. The articular ends of the centra are a rounded oval in outline, with the transverse diameter a little longer than the vertical—this difference being less in the anterior than in the posterior cervicals: there is a slight flattening beneath the neural canal. The articular faces are gently concave and are surrounded by a sharply defined rounded border;

---

[*] "On *Murænosaurus leedsi*, a Plesiosaurian from the Oxford Clay," Quart. Journ. Geol. Soc. vol. xxx. (1874) p. 200.

the centre is usually marked by a small dimple or pit indicating the spot where the notochord was originally situated. The sides of the centra beneath the bases of the neural arch are concave both from before backwards and from above downwards. The ventro-lateral regions of the centra are occupied by the rather prominent facets ($r.f.$) for the cervical ribs. The shape of these surfaces differs in the different species: in *Muraenosaurus leedsi* there is a prominent longitudinal ridge or fold immediately above

Text-fig. 51.

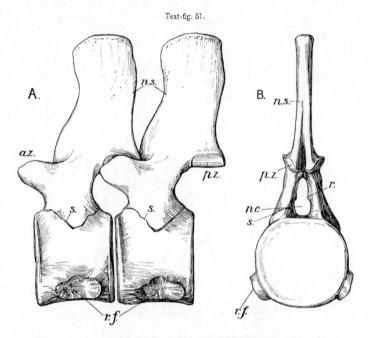

Posterior cervical vertebræ (? thirty-fourth and thirty-fifth) of *Muraenosaurus durobrivensis*: A, from left side; B, from back. (R. 2863, ¾ nat. size.)

a.z., anterior zygapophysis; *n.c.*, neural canal; *n.s.*, neural spine; *p.z.*, posterior zygapophysis; *r.*, ridge on inner face of neural arch; *r.f.*, facet for cervical rib; *s.*, neuro-central suture.

the rib-facet (Pl. IV. figs. 2-4). At about the fortieth vertebra the rib-facets begin to rise on the sides of the centrum, and on the forty-fourth and the next one or two vertebræ the facets are borne partly on the arch and partly on the centrum; the vertebræ on which this occurs are sometimes called pectorals (text-fig. 52). The ventral face of the cervical centra is concave from before backwards, and nearly flat or a little convex from side to side; on the middle line there is a low longitudinal ridge separating a pair of nutritive foramina. The edges of the centrum just before they

Text-fig. 52.

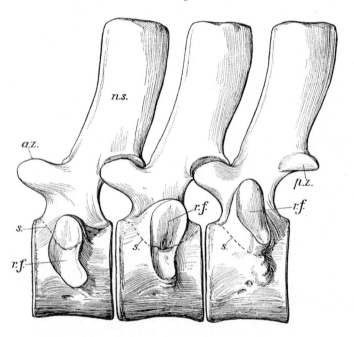

Pectoral and first dorsal vertebræ of *Murænosaurus durobrivensis*. (R. 2863, ⅔ nat. size.)
a.z., anterior zygapophysis; *n.s.*, neural spine; *p.z.*, posterior zygapophysis; *r.f.*, facets for ribs;
s., neuro-central suture.

pass into the articular faces are often marked by a series of fine plications, the presence of which seems to have led Phillips * to confer the name *Plesiosaurus plicatus* on the first named species belonging to the subsequently established genus *Murænosaurus*.

* 'Geology of Oxford' (1871) p. 313.

The pedicles of the *neural arches* unite with large diamond-shaped facets which extend nearly the whole length of the centra. In the cervical region, as in the rest of the vertebral column, the neural arch may or may not fuse with the centrum, its freedom or otherwise being apparently dependent entirely on the age of the individual. In the type specimen of *Murænosaurus leedsi* the fusion is complete, while in a much larger skeleton, to which Lydekker gave the name *Cimoliosaurus durobrivensis*, the arches are all free (text-figs. 50–52), so that probably when full-grown this species attained a much greater size than *Murænosaurus leedsi*. Above their base the pedicles narrow somewhat, but then widen out to the zygapophyses. Of these the anterior (*a.z.*) are roughly oval in outline and their articular faces are flat or, in the hinder part of the neck, concave from side to side; as usual they look upwards and inwards; the posterior (*p z.*) are similar in outline and are flat or somewhat convex from side to side, and look outwards and downwards. Both the anterior and posterior zygapophyses project considerably beyond the level of the articular ends of the centrum; the posterior to the greater extent. In about the anterior nineteen vertebræ a ridge runs backwards from the outer edge of the anterior zygapophysis (text-fig. 50, A) and posteriorly becomes continuous with the outer border of the posterior zygapophysis; in this region the neural spine (*n.s.*) rises from the middle of the surface bounded by these ridges, which in the posterior cervicals becomes less and less developed till they disappear, and the outer surfaces of the pedicles pass without interruption into the sides of the neural spine. In the anterior portion of the cervical region (text-fig. 50) the strongly compressed neural spines (*n.s.*) are low and confined to the posterior half of the arch, but further back they become very high (Pl. VI. fig. 4; text-fig. 51) and widen out towards their base, so as to extend forwards between the anterior zygapophyses; their anterior border is concave, the posterior convex, and their summits, which are abruptly truncated about at right angles to their long axis, are occupied by roughened surfaces, which in life were no doubt tipped with cartilage. In the pectoral vertebræ (text-fig. 52) the centra and the neural arches are, on the whole, similar to the posterior cervicals, except that the rib-articulation has passed upwards and is situated in part on the centrum and in part on a short stout transverse process arising from the lower part of the pedicle. This process becomes more and more prominent and rises on the arch till on about the 46th vertebra, which may be regarded as the first dorsal, where it alone bears the articulation for the rib.

The articular surfaces of the centra of the dorsal vertebræ (Pl. IV. fig. 5; Pl. V. fig. 5; text-figs. 53–55) are nearly circular in outline, there being only a slight concavity beneath the neural canal; they are nearly flat, with a small papilla having a pit at its summit, in their centre, and they are separated from the lateral and ventral surfaces of the centrum by a sharp and clearly-defined edge. The ventral surface is smooth and evenly convex from side to side; the lateral surfaces beneath the facets for union with the neural arch are somewhat concave and are perforated by large nutritive foramina.

The facets for the bases of the neural arch are diamond-shaped: they extend almost from end to end of the centrum and are deeply excavated. The floor of the neural canal is gently concave from side to side; it is greatly narrowed in the middle by the encroachment of the surfaces for the neural arch; near its middle there is a pair of small nutritive foramina.

Text-fig. 53.

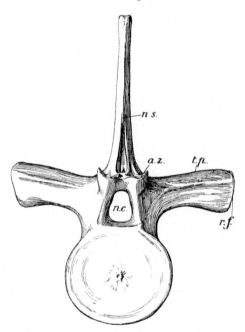

Anterior dorsal vertebra of *Murænosaurus durobrivensis*, from front. (R. 2863, ⅔ nat. size.)
a.z., anterior zygapophysis; *n.c.*, neural canal; *n.s.*, neural spine; *r.f.*, facet for rib;
t.p., transverse process.

In the dorsal region the pedicles of the neural arches are stout and, as already stated, unite with the centrum by a diamond-shaped surface, the suture (*s.*) becoming obliterated in old animals. The neural canal (*n.c.*) is oval in outline, the long axis being vertical. The neural spine (*n.s.*) is high and compressed; its border is slightly concave anteriorly and convex posteriorly, the spine as a whole, on most of the vertebræ, sloping backwards at its base and then turning to a vertical direction; its upper end is abruptly truncated by a surface for cartilage. Near the base of the spine on both the anterior and posterior edges there is a deep vertical groove for the attachment of ligaments; on the posterior surface this groove separates the posterior

Text-fig. 54.

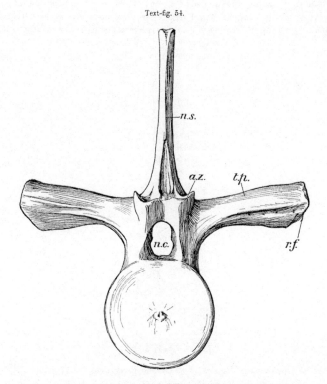

Middle dorsal vertebra of *Murænosaurus durobrivensis*, from front. (R. 2863, ⅔ nat. size.)
a.z., anterior zygapophysis; *n.c.*, neural canal; *n.s.*, neural spine; *r.f.*, facet for rib; *t.p.*, transverse process.

zygapophyses (*p.z.*). These form strong prominences and their articular surfaces, which are convex from side to side, look nearly directly downwards. The anterior zygapophyses (*a.z.*) are almost parallel with one another and separated anteriorly by a deep notch : their articular surface is concave from side to side. As already mentioned, in the two or three vertebræ (pectorals) behind the cervical series the rib-articulation is borne partly on the centrum and partly on the arch ; in the dorsal region it passes entirely on to the arch and is situated at the end of a more or less elongated transverse process (*t.p.*). In the anterior dorsal region these processes are comparatively short and stout (Pl. IV. fig. 5 ; text-fig. 53) ; they arise from the bases of the arches and are

Text-fig. 55.

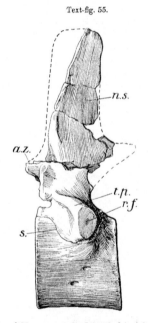

Posterior dorsal vertebra of *Murænosaurus durobrivensis*, from left side. (R. 2863, ⅔ nat. size.)
a.z., anterior zygapophysis ; *n.s.*, neural spine ; *r.f.*, facet for rib ; *s.*, neuro-central suture ;
t.p., transverse process.

somewhat curved, the concavity being downwards ; towards their extremities they enlarge and bear large convex facets (*r.f.*), for union with the single-headed ribs. Further back the processes arise higher and higher up the arch, and at the same time they become more slender at the base and curve a little backwards, so that the rib-facet looks less directly outwards than in the anterior dorsals; the anterior face of the processes is marked by a strong longitudinal ridge for muscle-attachment (text-fig. 54). In the posterior dorsals (text-fig. 55) the transverse processes rapidly shorten and pass down the arch to the lower end of the pedicle; behind this in the sacral and caudal regions no transverse processes are present, the ribs articulating directly with facets at the sides of the centra.

There are three or four sacral vertebræ (text-fig. 56), the centra of which become more and more depressed, so that the transverse diameter of the oval articular end is greater than the vertical diameter; at the same time they become shortened, a

Text-fig. 56.

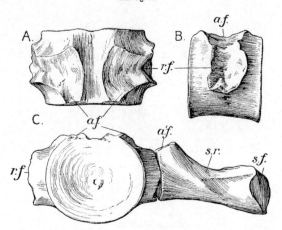

Sacral vertebra and rib of *Murænosaurus durobrivensis*: A, from above; B, from left side; C, from front with rib. (R. 2863, ⅔ nat. size.)

a.f., facet on centrum for neural arch; *a'.f.*, facet for neural arch on proximal end of sacral rib; *r.f.*, facet for sacral rib; *s.r.*, sacral rib; *s.f.*, facet for attachment with ilium.

process continued into the caudal series. The articular faces of the centra are nearly flat with a central tubercle. The ventral face is evenly and gently convex from side to side and is perforated laterally by a pair of foramina; externally its middle portion passes into the ventral surface of the prominence, which bears the deeply-cupped surface (*r.f.*) for union with the sacral rib (*s.r.*); this surface is situated a little nearer the posterior than the anterior end of the centrum and is continuous above with that for union with the neural arch (*a.f.*); this surface also is deeply concave and extends to, or nearly to, the posterior end of the centrum, but is separated from the anterior end by a considerable interval. The floor of the neural canal is gently concave from side to side; it is much narrowed between the surfaces for the neural arch, and widens out considerably in front of and behind them. The neural arches of the sacral vertebræ are not well known, but it can be seen that the outer side of the pedicle had a small surface for union with the upper side of the inner end of the sacral rib ($a'.f.$).

In the caudal region the centra of the vertebræ are shorter than in other parts of the column (Pl. IV. fig. 6; Pl. V. figs. 6, 7; text-figs. 57–59). At the same time they become much wider than high, and in consequence of the presence of large facets for

Text-fig. 57.

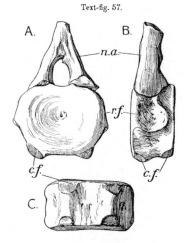

Middle caudal vertebra of *Murænosaurus durobrivensis*: A, from front; B, from left side; C, from below. (R. 2863, $\frac{2}{3}$ nat. size.)

c.f., facets for chevrons; *n.a.*, neural arch; *r.f.*, facets for caudal ribs.

the union of the arches and ribs they often present a more or less pentangular outline. The articular faces are rather more concave than in the dorsal and sacral regions. The facet for union with the arch is large, extending nearly the whole length of the centrum; the floor of the neural canal is concave from side to side and widest behind. The surface (*r.f.*) for union with the caudal rib is large and nearly circular; above it is usually continuous with that for the neural arch. Beneath it the sides of the centrum are slightly concave from above downwards, and are separated from the nearly flat ventral face by rounded ridges, the posterior ends of which in the anterior caudals, and both the anterior and posterior ends in the middle and the posterior caudals, are truncated by oblique facets (*c.f.*) for union with the chevrons; when both anterior and posterior facets are present, the latter are usually the larger (text-figs. 57-59, *c.f.*). At the posterior end of the caudal region (text-fig. 58) the centra become reduced in size very rapidly; the actual terminal vertebræ are not known in this genus. In the anterior part of the caudal region the neural arches are not well known, but it appears that they possessed well-developed zygapophyses and comparatively high laterally compressed neural spines. Further back beyond the middle of the tail (text-figs. 58, 59) the

Text-fig. 58.

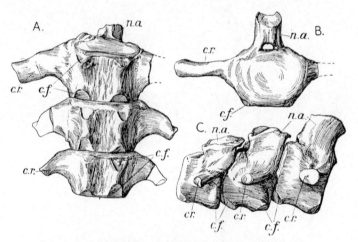

Posterior caudal vertebræ of *Murænosaurus platyclis*: A, from below; B, from front; C, from right side. (R. 2425, ¾ nat. size.)

c.f., facets for chevrons; *c.r.*, caudal ribs; *n.a.*, neural arch.

arch becomes smaller and stouter, bearing a short thick neural spine (*n.s., n.a.*) terminating above in a surface for cartilage; in this region the zygapophyses have disappeared, or are represented only by slight rugosities. Further back still the arch is more massive and encloses a very small neural canal. The spine is very short and thick, and is almost rectangular in section; it slopes a little backwards and terminates above in a large, slightly concave surface for a cap of cartilage. There is no trace of

104   MARINE REPTILES OF THE OXFORD CLAY.

any flexure in the caudal region such as is seen in *Ophthalmosaurus*. The chevrons (text-fig. 59) are paired compressed rods of bone which do not unite with one another ventrally; they are usually slightly curved, at least in the anterior part of the tail, where they unite with single facets on or near the hinder border of the

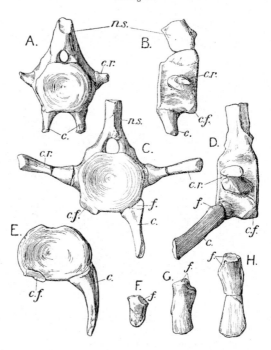

Text-fig. 59.

Caudal vertebræ and chevrons of *Muræenosaurus leedsi*: A, posterior caudal from back; B, same specimen from right side; C, middle caudal vertebra from back; D, same specimen from right side; E, anterior caudal from back; F, G, H, three chevron-bones from side. (R. 2864, ⅔ nat. size.)

c., chevron-bones; *c.f.*, facets for chevron-bones; *c.r.*, caudal ribs; *f.*, facets of chevron-bones for articulation with caudal centra; *n.s.*, neural spine.

centra (*c.*, text-fig. 59, E). Further back they articulate by two facets (*c.f.*) between the centra of successive vertebræ; the two facets make an obtuse angle with each other, and the anterior one is usually the larger; as these bones are followed towards the end of the tail they become shorter and shorter (text-fig. 59, H, G, F), disappearing entirely in the last few vertebræ.

The cervical *ribs* (r., Pl. IV. figs. 2, 3, 4 ; Pl. VI. figs. 4, 5) are all single-headed, the only sign that they may at one time have possessed two heads being perhaps shown in the peculiar form of the rib-facets in some of the cervical vertebræ of *Murænosaurus durobrivensis*, where the articular surface is produced a little upwards into a sort of small dorsal lobe (text-fig. 50, A, $r.f.$), which, however, is quite continuous with the remainder of the surface.

On the atlas the rib ($r.^1$, text-fig. 49) is merely a small pointed process, but on the axis ($r.^2$) it is much larger and similar to the succeeding cervical ribs. These vary in form considerably in the different species. In *Murænosaurus leedsi* (Pl. IV. figs. 2–4) they are strongly compressed from above downwards, and on the anterior portion of the neck have a well-marked anterior angulation of the blade, thus showing some approximation to the hatchet-like form found in many Liassic Plesiosaurs and in *Picrocleidus beloclis*. In *M. platyclis* the ribs are likewise greatly compressed from above downwards into thin plates; but the anterior angulation is well marked only in the anterior vertebræ in some specimens, and at the back of the neck they become broad, thin, and slightly backwardly directed plates of bone (Pl. VI. fig. 4). In both these species there is a longitudinal ridge on the side of the centrum immediately above the facet for the rib (Pl. IV. figs. 2, 3). In *Murænosaurus durobrivensis* the cervical ribs are much less compressed than in the other species, especially in the posterior part of the neck, where they are oval in section, at least near the base.

The dorsal ribs (text-fig. 60, C, D) articulate with the ends of the transverse processes by large slightly concave oval surfaces, the long diameter of which is vertical. Towards the articulation the bone becomes somewhat compressed from before backwards, and its surface bears strong ridges for the attachment of muscles. Further out it becomes rounded in section and terminates abruptly at its ventral end in a concave surface, indicating that in life it was tipped with cartilage.

There were either three or four pairs of sacral ribs (text-fig. 56, $s.r.$), which differ from one another considerably in form, and their arrangement with regard to one another is not certain. They are all stout bars of bone, enlarged at their extremities, the inner end of each bearing a strongly convex surface for union with its centrum, with which it may fuse; the upper border of the proximal end may also bear a facet ($a'.f.$) looking upwards and articulating with the outer part of the pedicle of the neural arch. The outer end terminates in a convex facet ($s.f.$), which no doubt was connected in some way with the ilium. In some of the sacral ribs the shaft is curved, so that probably they converged towards their outer ends.

The caudal ribs (text-figs. 58, 59, $c.r.$) articulate with the centra by strongly convex facets; they are compressed from above downwards, sometimes to a high degree, and in the posterior portion of the tail are curved backwards, their outer ends being occupied by a nearly flat facet.

In both the cervical and caudal regions the ribs may fuse with the centra or remain separate. Their freedom or fusion is not of any systematic significance, except so far as it may indicate that in some forms the adult condition is attained while the animal is small, while in others the fusion does not take place till it has reached a large size. The same remarks hold good with regard to the fusion or separation of the neural arches throughout the column.

The arrangement of the ventral ribs (text-fig. 60, A, B) is not well known in this genus: probably they were placed much as in *Cryptocleidus* (see below, p. 175)—at least the individual elements are closely similar to those found in that genus, in which

Text-fig. 60.

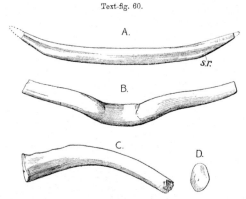

Ventral and dorsal ribs of *Muraenosaurus durobrivensis*: A, median ventral rib; B, posterior median ventral rib; C, dorsal rib from behind; D, articular end of dorsal rib. (R. 2863, ¼ nat. size.)
s.r., surface for union with lateral ventral rib.

each of the transverse rows (with the possible exception of one or two posteriorly) consists of a median and three pairs of lateral bones. The median bone (text-fig. 60, A, B) is overlapped at either end by the inner ends of the first lateral pair, which are closely applied to its anterior face: these again are overlapped in a similar way by the second pair, and these again by the inner ends of the outer pair (see figure of plastron of *Cryptocleidus*, text-fig. 86, p. 175).

Shoulder-girdle (Pl. VI. figs. 3, 6; text-figs. 61, 62, 67, 68).—Before describing the structure of the shoulder-girdle in *Muraenosaurus* it may be well to give a brief account of this portion of the skeleton in the Sauropterygia generally, especially as its structure has been the subject of much controversy and can only be understood by reference to the earlier members of the group, such as the Triassic Nothosaurs. In these reptiles the shoulder-girdle consists of scapulæ, enlarged coracoids, and a more or less complete arch of bones forming an anterior transverse bar, of which the outer ends are united to ventral processes of the scapulæ. The constituent elements of this bar are three in number, one median and two lateral, and it is concerning the nature of these bones that discussion has arisen, some anatomists regarding them, or at any rate the median one, as of sternal origin, others as clavicles and interclavicle and therefore as membrane-bones. This latter interpretation has been strongly, and, in my opinion, rightly, supported by the late Professor Seeley, and, as in my former papers, this view is here adopted. The presence of this clavicular arch closely united at its outer ends with the scapulæ, seems to be correlated with the adoption of an aquatic life and the consequent alteration in the direction of the pressure on the shoulder-girdle through the glenoid cavity; since under the new conditions instead of the fore limbs helping to support the weight of the body, as they do in most terrestrial animals, they now are mainly concerned in propelling it forwards through the water. The consequence of this is, that the thrust of the head of the humerus on the glenoid cavity is directed more or less forwards instead of mainly upwards, and the presence of the transverse clavicular bar attached to the scapulæ is necessary to resist the forward and inward thrust; and from the Nothosauridæ through the Plesiosauridæ to the most highly specialised Elasmosauridæ the modifications of the shoulder-girdle that have taken place are in the direction of increased rigidity in that region. In the course of this series of changes there is a tendency to the reduction of the clavicular arch accompanied by, and consequent upon, the increasing size and importance of the ventral ramus of the scapula. In the Triassic genus *Nothosaurus* the clavicular arch (text-fig. 61, B) is a long curved bar composed of a pair of elongated clavicles, uniting at their median ends with one another and with the small interclavicle, and externally uniting in suture with a small ventral expansion of the scapulæ. In most of the Liassic Plesiosaurs (text-fig. 61, A) the ventral processes of the scapulæ are increased in size, and in some cases may be in contact with one another in the middle line. The clavicular arch, which is overlapped below by the ventral processes of the scapulæ, is modified in various ways: as a rule, the interclavicle is large and extends back to the coracoids, so that the space between the clavicular arch and the coracoids is divided into two separate openings; the clavicles may be much reduced. In the animal which has been called by Lydekker[*] *Thaumatosaurus arcuatus* the condition is rather different, the clavicle and interclavicle being both large and the latter not extending back to the coracoids, but having a concave posterior border. In *Murænosaurus* (text-figs. 62, 68), *Cryptocleidus* (text-figs. 87–89), and other members of the Elasmosauridæ the rigidity

[*] Catal. Foss. Rept. Brit. Mus. pt. ii. (1889) p. 163.

of the anterior part of the shoulder-girdle is mainly effected by the great increase in size of the ventral bars of the scapulæ, which, in the adult, meet in median symphysis and also send back processes which meet the median anterior prolongations of the coracoids. The consequence of this extension of the scapulæ inwards to the middle line is, that the clavicular arch comes to lie on the visceral surface of those bones and becomes functionless, at least in the adult animal, and undergoes reduction. This may take place in several ways, and there is great variability in the degree to which it is carried, practically all stages from the presence of a well-developed clavicle and

Text-fig. 61.

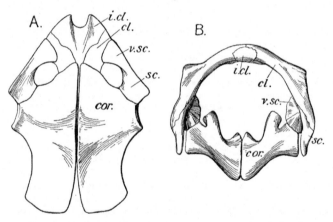

Shoulder-girdle of: A, *Plesiosaurus ? rostratus* (R. 1315, about ¼ nat. size), from below; B, *Nothosaurus* sp., from above. (About ¼ nat. size.)

cl., clavicle; *cor.*, coracoid; *i.cl.*, interclavicle; *sc.*, scapula; *v.sc.*, ventral ramus of scapula.

interclavicle to the almost complete absence of either or both these elements. Some of the various forms are shown in the text-figures of the shoulder-girdle in the different genera.

In *Murænosaurus* (Pl. VI. fig. 3; text-figs. 62, 67, 68) the *clavicles* in most cases undergo great reduction and may become mere paper-thin plates of bone adherent to the inner face of the interclavicle (Pl. VI. fig. 3) and of the ventral ramus of the scapula. The *interclavicle* (Pl. VI. fig. 6) is usually a well-developed, more or less oval disc of bone, thin at the margin, but thickening towards the middle; its anterior border bears in the middle line a rounded notch with somewhat thickened edges, while the middle of the posterior border may either bear a similar rounded notch (*M. durobrivensis*, Pl. V. fig. 10; *M. platyclis*, Pl. VI. fig. 3; text-fig. 68) or it may be produced back into a short pointed process (? *M. leedsi*, Pl. VI. fig. 6; text-fig. 62); in the former case the notch may form the anterior border of the interscapular foramen (*i.s.f.*, Pl. VI. fig. 3; text-fig. 68). The variations of the shoulder-girdle in this genus will be noticed under the different species into which it has been found desirable to distribute it.

Text-fig. 62.

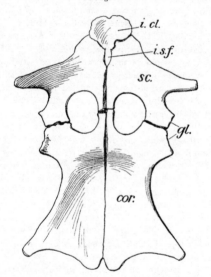

Shoulder-girdle of *Murænosaurus* ? *leedsi*. (R. 3704, ⅓ nat. size.)

cor., coracoid; *gl.*, glenoid cavity; *i.cl.*, interclavicle; *i.s.f.*, interscapular foramen; *sc.*, scapula.

The *scapulæ* (*sc.*, Pl. VI. fig. 3; text-figs. 62, 67, 68), as in the other genera of the Elasmosauridæ, are triradiate bones, each consisting of a posterior arm uniting with the antero-external angle of the coracoid, a dorsal arm extending towards the vertebral column, and a ventral expansion which, in the adult animal, meets its fellow in the middle line in a deep symphysial surface, which extends back to meet the anterior median prolongations of the coracoids. The posterior bar is thickened at its extremity, where it bears two facets, one for union with the coracoid, the other forming the anterior part of the glenoid cavity. The first of these facets looks backwards and inwards; it is triangular in outline and its nearly flat surface bears rugosities which indicate that in life it was capped with cartilage. The glenoid facet looks backwards and outwards, making an angle of about 90° with that for the coracoid; in outline it is half an oval, the line of union with the other surface being the short diameter of the oval; it is only slightly concave and nearly smooth in specimens in which ossification seems to be approaching completion, but in younger animals it may be roughened, and was no doubt covered with cartilage. The shaft of the posterior bar is almost triangular in section: one angle forms the sharp outer border of the coracoscapular fenestra; the lower angle forms a strong ridge which extends on the outer (lower) face of the bone from the anterior angle of the glenoid cavity to the anterior angle of the blade, forming the line of division between the dorsal and ventral portions. The dorsal arm is directed a little backwards; it rises from a broad base, but narrows above, the anterior and posterior borders becoming nearly parallel; its upper suprascapular border is terminated by a narrow concave surface, indicating the presence of a cap of cartilage. The size and shape of the ventral ramus of the scapula vary both according to the species and with the age of the individual. Probably in all specimens of fully grown animals the scapulæ of opposite sides meet in median suture and extend back to meet the anterior median prolongations of the coracoids. In the type specimen of *M. platyclis*, in which growth was complete, the scapula (Pl. VI. fig. 3; text-fig. 68) has a broad ventral plate, of which the concave upper surface is continuous with the upper surface of the dorsal ramus, but the ventral face is sharply separated from the outer face by the ridge-like angle referred to above. The ventral plates of the opposite scapulæ meet in median suture in their posterior half only, where they are much thickened; the thin anterior portions are separated by a deep notch, the posterior end of which formed the posterior boundary of the oval interscapular foramen, the anterior border of which is constituted by the posterior notch of the interclavicle (*i.s.f.*, Pl. VI. fig. 3; text-fig. 68). The posterior prolongations towards the coracoid are nearly semicircular in section, the diameter of the semicircle being the sutural surface by which they unite with one another; posteriorly they join the anterior bars of the coracoids by semicircular surfaces which, in some cases, are further forward on one side than the other, owing to the anterior prolongation of one coracoid being longer than that of the other (Pl. VI. fig. 3). Except for a slight prominence on its outer third, the anterior border of the ventral ramus is nearly straight, and it runs inwards and backwards towards the middle line, the anterior border of the united scapulæ forming a widely open **V**, which in life was covered by the interclavicle, which only rested on the inner face of the scapulæ by its outer and posterior borders, and was not entirely shut in by bone below, as are the clavicles in adult specimens of *Cryptocleidus*; possibly, however, the **V** was partly closed by cartilaginous extensions of the edges of the scapulæ. The arrangement of the interscapular foramen has been referred to above.

The *coracoids* (text-figs. 62, 67, 68) are very large and broad plates of bone, thin posteriorly, with a much thickened region between the glenoid cavities, the thickening being indicated on the visceral surface by a strong antero-posterior convexity. Internally the two bones unite with one another in the middle line in a long suture, which, in the adult, is continued forwards between the anterior median prolongations of the bones till it becomes continuous with the symphysis of the scapulæ. In its posterior portion the symphysial surface is thin, but at the thickened portion of the bone it widens out, again narrowing a little between the anterior prolongations, which are approximately semicircular in section like the posterior processes of the scapulæ. The form of the posterior border of the bone is best understood from the figures. The postero-lateral processes do not seem to develop till late in life (*cf.* text-fig. 62) and never attain the strong development found in the coracoids of *Cryptocleidus*. The outer border is evenly concave and towards the middle is sharp-edged, thickening anteriorly and posteriorly. Anteriorly it terminates on the hinder angle of the broad glenoid surface which looks outwards and forwards and is half an oval in outline, the line of junction with the glenoid surface of the scapula being the short diameter of the oval; this surface is very slightly concave, and in large individuals nearly smooth. The facet for articulation with the head of the scapula looks outwards and forwards, making an angle of about 135° with the glenoid surface; it is triangular and much roughened. The anterior edge of the bone forming the posterior border of the coraco-scapular opening is thin and sharp.

Fore Limb (Pl. IV. fig. 7; Pl. V. fig. 11; text-fig. 63, B).—The fore limb is considerably larger than the hind limb[*], and the distal end of the humerus is more expanded than that of the femur. In older individuals of the larger species (*M. durobrivensis*, Pl. V. fig. 11) the humerus may become extremely massive and the distal end proportionately wider, although there is no approach to the great expansion seen in the humerus of *Cryptocleidus*. The upper end is considerably enlarged and bears the head and the great tuberosity, which is situated on the dorsal surface a little towards the posterior side. The head does not seem to have ever become fully ossified, the surface even in the oldest individuals being only slightly convex and covered with rugosities and tubercles, which seem to indicate that a considerable capping of cartilage was present, a conclusion which is supported by the form of the glenoid cavity; probably the cartilage-covered head was nearly hemispherical. Above and posteriorly the cartilage-covered surface of the head is continuous with that forming the upper end of the great tuberosity, the two surfaces making a slight angle with one another. The upper end of the tuberosity is roughly oblong in outline, its outer angle forming strong prominences; from the

---

[*] In the original description of *Murænosaurus leedsi* it is stated that the fore limb is the smaller, but the paddle described is really the hind limb (Quart. Journ. Geol. Soc. vol. xxx. (1874) p. 207).

112   MARINE REPTILES OF THE OXFORD CLAY.

hinder angle a short crest runs down on to the hinder face of the upper part of the shaft. Both the upper end of the tuberosity and of this ridge seem to have been

Text-fig. 63.

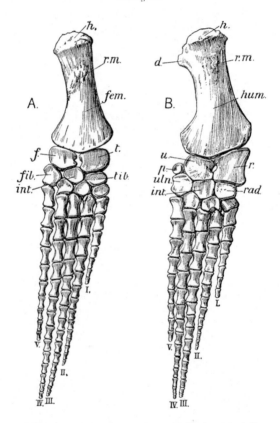

Paddles of *Murænosaurus leedsi*: A, left hind paddle from ventral side
B, left fore paddle from ventral side. (R. 2864, about ⅕ nat. size.)

d., tuberosity of humerus; *f.*, fibula; *fem.*, femur; *fib.*, fibulare; *h.*, head of humerus and femur; *hum.*, humerus; *int.*, intermedium; *p.*, pisiform; *r.*, radius; *rad.*, radiale; *r.m.*, ridges for muscle-attachment; *t.*, tibia; *tib.*, tibiale; *u.*, ulna; *uln.*, ulnare; I.-V., first to fifth digits.

covered with cartilage in life. The shaft is somewhat compressed from above downwards, its narrowest part being about the middle of its length, where it is oval in section. On the ventral face near the preaxial border and a little above the middle of the shaft is a strongly marked rugose surface for the attachment of muscles; in old individuals much of the upper part of the ventral face of the bone, both above and below this surface, is considerably roughened by ridges and tubercles. On both the ulnar and radial borders of the shaft there are smaller rugosities for muscle-attachment. Both the anterior and posterior borders of the bone are concave, the posterior concavity being the most marked. The distal end is compressed dorso-ventrally and much expanded, especially in old individuals in which ossification is far advanced, but, as mentioned above, the expansion is never so great as in the humerus of *Cryptocleidus*. At its distal end the humerus articulates with the radius and ulna, the surfaces for which meet one another at a very obtuse angle. These surfaces are rough and in life were no doubt covered with cartilage, which also extended along the postaxial and to some degree the preaxial borders of the expansion; it is possible that small ossifications may have occurred in the pre-radial or more often in the post-ulnar cartilage. Such an ossification is shown in the figure of the fore limb of *Tricleidus* (text-fig. 77). The surface of the humerus has a peculiar fibrous appearance, the direction of the "fibres" being longitudinal. The middle of the shaft is nearly smooth, but towards the extremities, especially on the distal expansion, the surface is pitted by a number of small foramina running inwards towards the middle of the shaft.

The general form of the *radius* (r., Pl. IV. fig. 7; text-fig. 63, B) will be best understood from the figures. The oblique proximal surface for union with the humerus is gently convex, the comparatively thin anterior (preaxial) border is gently concave, while the thickened ulnar (postaxial) border is sharply notched. Distally the bone articulates with the intermedium by a short facet looking downwards and backwards, and with the radiale by a gently concave or flat surface. The *ulna* (u., Pl. IV. fig. 7; text-fig. 63, B) is much smaller than the radius, particularly in length; its anterior (radial) border is deeply notched, and the notch with the corresponding one on the radius encloses a foramen, representing the opening between the shafts of the two bones before they had become so much shortened. The posterior border is convex, while the distal bears two facets making a very obtuse angle with one another; the anterior facet, which is the larger, articulates with the intermedium, the posterior with the ulnare.

The proximal row of *carpals* (text-fig. 63, B) consists in most cases of three elements only, but sometimes there is a fourth, which is a small bone articulating with the postero-external angle of the ulna and the postaxial portion of the proximal face of the ulnare. This element may be a pisiform (p., text-fig. 63, B), and is probably homologous with the third element which in some cases articulates with the humerus, *e.g.*, in *Tricleidus* and *Dolichorhynchops* and *Polycotylus*. Williston regards this ossicle and others similarly situated, as new ossifications in the postaxial cartilage, and in many cases he is probably right; he proposes the name "first epipodial supernumerary" for the bone here regarded as pisiform. All the three usual bones of the proximal row of carpals are flat polygonal elements, which, when fully ossified, fit closely together as well as against the radius and ulna and the distal carpals, making with them a close pavement. The *radiale* (*rad.*) unites with the radius, intermedium (*int.*), and the first and second distal carpals. The *intermedium* (*int.*) has a small surface for union with the radius and a long one for the ulna; it also joins the radiale, ulnare, and the second and third distal carpals. The *ulnare* (*uln.*) unites proximally with the ulna and pisiform (if present); it also articulates with the intermedium, the third distal carpal, and the fifth metacarpal. The distal carpals appear to be only three in

Text-fig. 64.

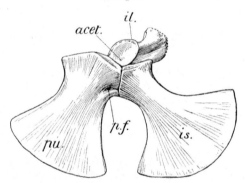

Pelvis of *Nothosaurus* sp., from left side, showing the union of the ilium, ischium, and pubis in a triradiate suture to form the acetabulum. (38675, ½ nat. size.)

acet., acetabulum; *il.*, ilium; *is.*, ischium; *p.f.*, pubic foramen; *pu.*, pubis.

number, the fifth metacarpal, as already noted, articulating directly with the ulnare. They are polygonal bones, somewhat smaller than the proximal row; the first carries the first digit only, the second the second and third digits, although in some cases there may be a slight contact also with the first. The third distal carpal carries the fourth digit, and on its outer (postaxial) side it has a surface for contact with the preaxial face of the metacarpal of the fifth digit.

The *metacarpals* in some cases are flattened like the carpals, but, as a rule, are somewhat cylindrical and larger at the ends than in the middle. In addition to their surfaces for union with the carpals they often have at their proximal ends small lateral facets for contact with one another. The phalanges are very numerous: they are cylindrical and much constricted in the middle; they become smaller and smaller towards the finger-tips, where they may be represented by mere nodules of bone. In the most nearly perfect fore paddle preserved (R. 2864, text-fig. 63, B) the numbers of phalanges in the digits I.–V. are 6, 12, 14, 13, 8. In no specimen has the presence of more than the normal five digits been observed.

Pelvis (Pl. IV. figs. 8, 9; Pl. V. figs. 8, 9; text-fig. 65).—The pelvis is composed of the usual three pairs of bones, and like that of the other Plesiosaurs is chiefly remarkable from the fact, that the three elements of each side do not meet in the acetabulum in a triradiate suture, the ilium having been rotated backwards and its lower end having lost its connection with the pubis. In the early Sauropterygia (text-fig. 64) the pelvis is in this respect of normal type and the peculiarity of this part of the skeleton in the later forms, as in the case of the shoulder-girdle, seems to be the result of the different mechanical conditions brought about by the adoption of an entirely aquatic life. The *ilium* (Pl. IV. figs. 8, 8 a; Pl. V. figs. 8, 8 a, 8 b; *il.*, text-fig. 65) is a curved rod of bone expanded towards its extremities. The upper end is compressed from within outwards and slightly everted; the upper end (*cr.i.*) is obliquely truncated by a surface which was tipped with cartilage in life. The inner, slightly convex face of this expanded upper end is sometimes roughened, as if for union with the distal end of the sacral ribs (*s.f.*, Pl. IV. fig. 8 a; Pl. V. fig. 8), but in other cases is nearly smooth and shows no trace of such union: probably the age of the animal may have something to do with these differences. The shaft also is a little compressed from within out and is oval in section; it is somewhat curved, the concavity being anterior. On the convex posterior border there is a strong ridge forming a projecting angle at about the middle of the bone. The thickened lower end bears two facets—one large and slightly concave, looking forwards, downwards, and outwards for union with the ischium (*is.f.*); the other (*acet.*) small and continuous with the last, but making an angle with it and looking mainly forwards; this facet forms the posterior lip of the acetabular surface, and it is continued a little upwards on to the prominent antero-external angle of the distal expansion (Pl. V. figs. 8 a, 8 b, c.). Both surfaces were covered with cartilage in life.

The *pubis* (text-fig. 65, p.) is a large plate of bone, the form of which will be best understood from the figure. It is a little wider than long, and the greater part of it is very thin, but towards the outer and anterior borders and in the region of the symphysis it is thickened. The inner border by which the bone unites with its fellow of the opposite side in a median symphysis is straight or convex. The symphysial surface is deepest at about the middle of its length, where the bone is considerably thickened. The anterior border is at first straight or slightly undulated, then it curves outwards and backwards to the somewhat prominent antero-external angle where it joins the outer border; throughout its length this anterior edge is grooved and must have borne a

116   MARINE REPTILES OF THE OXFORD CLAY.

fringe of cartilage. The outer border is concave and terminates posteriorly in the anterior angle of the acetabular surface; its edge is sharp and was not fringed with cartilage. The posterior border is strongly concave and its edge is rather sharp; it

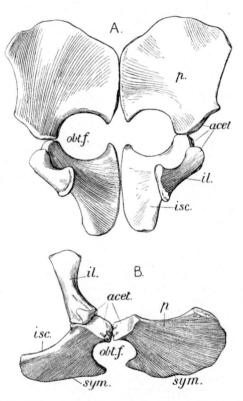

Pelvis of *Murænosaurus durobrivensis* : A, from above ; B, from right side.
(Type specimen, R. 2428, ½ nat. size.)

acet., acetabulum ; *il.*, ilium ; *isc.*, ischium ; *obt.f.*, obturator foramen ; *p.*, pubis ;
sym., symphysial border.

forms the anterior border of the obturator foramen (*obt.f.*). Between the posterior and lateral borders is the thickened and massive postero-external portion of the bone, which bears the acetabular and ischial surfaces. The former of these is much the larger: its form is that of half an elongated oval, the short diameter of which is the line of union with the ischial surface; it is gently concave in a longitudinal direction and forms the greater part of the surface for the articulation of the femur; there is no union with the ilium. The ischial surface is roughly triangular in outline, the angles being rounded; it makes an angle of about 145 degrees with the acetabular surface. Both of these surfaces are roughened and were cartilage-covered, but on the acetabular surface the roughness is slight.

The *ischium* (Pls. IV. & V. fig. 9; *isc.*, text-fig. 65) is shaped somewhat like the head of a hatchet, consisting of a massive articular portion united by a comparatively narrow neck with the broad ventral blade. The head bears three articular surfaces. The anterior one (*pu.f.*) looks directly forwards; it is roughly triangular and unites with the corresponding surface of the pubis. Behind this and making an angle of about 90° with it, is the acetabular surface (*acet.*), roughly quadrate in outline and looking directly outwards. Behind this again, and making a very obtuse angle with it, is the triangular surface for the ilium (*il.f.*), looking upwards and backwards. The acetabular surface is comparatively smooth, but the others are raised into strong ridges with deep pits between them. Beneath the massive head the bone is much narrowed, and at the same time is compressed from above downwards, passing into the broad ventral expansion. The anterior and posterior borders of the bone are both concave, the anterior more strongly so; the anterior is rounded, while the posterior is sharp in the region of the neck of the bone, but towards the posterior angle it becomes roughened for the attachment of muscles. The median border in its anterior (symphysial) part is nearly straight, and, owing to the thickening of the bone, the symphysial surface (*sym.*) is deep, narrowing both forwards and backwards. Behind the symphysis the inner border of the ischium curves sharply outwards, meeting the posterior border at an angle of about 90°; this portion of the inner border has a grooved edge, and was fringed with cartilage during life. The ventral surface of the blade of the bone is nearly flat, except where the symphysial thickening occurs; the upper surface is gently concave. The symphysis of the ischia is not continuous with that of the pubis, so that so far as the bones are concerned the obturator foramina of opposite sides are in free communication with one another, but probably in life they were separated by a median band of cartilage.

Hind Limb (Pl. IV. fig. 10; Pl. V. fig. 12; text-fig. 63, A).—The *femur* is in most respects very similar in structure to the humerus, the most notable difference being the smaller extent to which the distal end is expanded. The head (*h.*) seems never to become completely ossified, but even in old individuals is only moderately

Text-fig. 61.

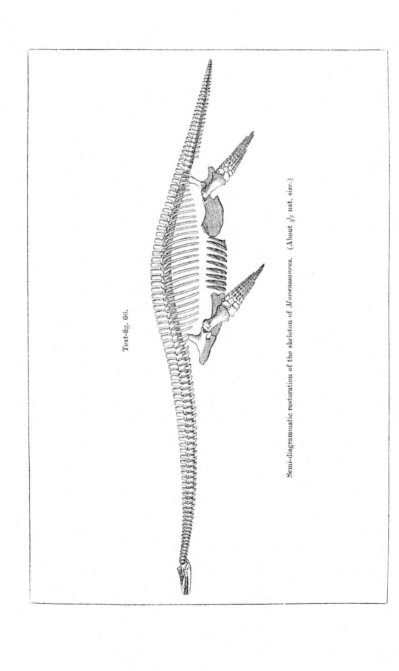

Semi-diagrammatic restoration of the skeleton of *Murænosaurus*. (About $\frac{1}{25}$ nat. size.)

convex and clearly had a thick cap of cartilage in life. The trochanter (*tr.*) arises from the dorsal surface just beneath the head, with which its upper end is continuous, the two making an angle of about 110° with one another. In outline the upper end of the trochanter is roughly quadrangular, but in some cases it is divided into a larger anterior and a smaller posterior portion by a deep groove running down its postero-superior face. The shaft is nearly circular in section, and the ventral surface of its upper half is much roughened for muscle-attachment (*r.m.*), and there is also a strong roughened ridge about the middle of the posterior border of the shaft. As already mentioned, the distal expansion is less than in the humerus, and the cartilage-covered terminal surface is not carried on to the postaxial and preaxial borders by the expansion, as in that bone. The surface of the bone has the same fibrous appearance noted in the case of the humerus.

Two bones only (the *tibia* and *fibula*) articulate with the distal end of the femur. Both are flattened polygonal bones, which articulate with one another by short proximal and distal surfaces, being separated in the middle by a foramen. Their surfaces for union with the femur are gently convex, so also to a greater degree are their outer borders. Distally the tibia (*t.*) bears a long facet for union with the *tibiale* (*tib.*) and a short one for the *intermedium* (*int.*), while the fibula (*f.*) joins the intermedium and *fibulare* (*fib.*) by facets of about equal length. The proximal tarsals are polygonal flattened bodies; the tibiale (*tib.*) unites distally with the first and second distal tarsals, the intermedium (*int.*) with the second and third, the fibulare (*fib.*) with the third and with the fifth metatarsal. The postaxial border of the fibulare is thin and concave. There are five *metatarsals*, the first borne exclusively by the first distal tarsal, the second by the first and second, the third by the second and third, the fourth by the third only, while the fifth, as already noted, articulates directly with the fibulare, just as the fifth metacarpal does with the ulnare. The metatarsals, especially the first, are somewhat flattened from above downwards, but the phalanges are cylindrical, with constrictions in the middle; the terminals may be mere nodules of bone. In the best paddle preserved (text-fig. 63, A) the numbers of the phalanges in the digits from the first to the fifth are 6, 13, 15, 13, 9, respectively.

Text-figure 66 is a semi-diagrammatic restoration of the skeleton of *Murænosaurus*, in which an attempt is made to represent the skeletal characters described above, and at the same time to give a general idea of the form of the animal as a whole. The chief points of interest are the relatively small size of the head, the great length of the neck, and the slight degree to which the distal end of the humerus is expanded compared with that of *Cryptocleidus*: the peculiar characters of the shoulder-girdle and pelvis cannot be shown in a profile view. The reconstruction is founded mainly on the skeleton of the type specimen of *Murænosaurus platyclis* (R. 2678).

Murænosaurus leedsi, Seeley.

[Plates III. & IV.; Plate VI. fig. 6; text-figs. 44, 59, 62, 63.]

1874. *Murænosaurus leedsi*, Seeley, Quart. Journ. Geol. Soc. vol. xxx. p. 197, pl. xxi.
1881. *Plesiosaurus leedsi*, Whidborne, Quart. Journ. Geol. Soc. vol. xxxvii. table facing p. 480.
1888. *Plesiosaurus plicatus*, Lydekker, Geol. Mag. [3] vol. v. p. 351.
1889. *Cimoliosaurus plicatus*, Lydekker, Catal. Foss. Rept. Brit. Mus. pt. ii. p. 234 (pars).
1895. *Murænosaurus plicatus*, Andrews, Ann. Mag. Nat. Hist. [6] vol. xvi. p. 429 (pars).

Type Specimen.—An imperfect skeleton, including portions of the skull and mandible, most of the vertebral column, ribs, portions of the pectoral and pelvic girdles, and the fore and hind paddles (R. 2421, Leeds Coll. 25), described and figured by Seeley, *loc. cit. supra* (see also Pls. III. & IV.; text-figs. 44, 59, 62, 63).

In 1889 Mr. Lydekker referred to this form as *Cimoliosaurus plicatus*, but subsequent descriptions * of the material from which *Cimoliosaurus* was defined by Leidy † show that this generic name cannot be employed in the wide sense in which Lydekker ‡ used it. The trivial name *plicatus* was used by Phillips § for a Plesiosaur which is undoubtedly a species of *Murænosaurus*, and was regarded as identical with *M. leedsi* by Lydekker ‖ and the present writer ¶; but since it is now found that several species of the genus occur in the Oxford Clay, and it is not certain to which of these the vertebræ described by Phillips belong, it seems best to reject his name altogether and adopt Seeley's name, his being the first description founded on adequate material.

This species does not appear to have reached the large size and massive proportions attained by *Murænosaurus durobrivensis*; at least the fusion of the cervical and caudal ribs and of the neural arches is complete in comparatively small individuals. It is, of course, impossible to say that these had reached their limit of growth, but it is certain that in the other species a much larger size was attained before these signs of complete ossification appeared. The same is true of the ossification of the scapulæ.

Unfortunately the pectoral and pelvic girdles are very imperfect in the type specimen, so that the important characters of these parts cannot be observed, but must be inferred from other specimens. The characters by which this species is distinguished from the other species here recognized, viz. *M. durobrivensis* and *M. platyclis*, are:—
(1) The relatively small size of individuals in which ossification seems to be

---

* Williston, "North American Plesiosaurs: *Elasmosaurus*, *Cimoliasaurus*, and *Polycotylus*," Amer. Journ. Sci. [4] vol. xxi. (1906) p. 221.
† Leidy, Proc. Acad. Nat. Sci. Philad. 1851, p. 325 (1852).
‡ Lydekker, Catal. Foss. Rept. Brit. Mus. pt. ii. (1889) p. 180.
§ Phillips, Geology of Oxford, etc. (1871) p. 313.
‖ Lydekker, *loc. cit. supra*.
¶ Andrews, "The Pectoral and Pelvic Girdles of *Murænosaurus plicatus*," Ann. Mag. Nat. Hist. [6] vol. xvi. (1895) p. 429.

approaching completion. (2) The greater length of the centra of the cervical vertebræ compared to their width, particularly in the hinder part of the neck; in this region also in this species the centra are longer in proportion to their height, and the articular ends are more oval than in *M. durobrivensis*. (3) The presence of a strong longitudinal crest on the side of the cervical centra just above the rib-joint distinguishes this species from *M. durobrivensis* but not from *M. platyclis*. (4) The interclavicle, in some cases at least, is produced back in the middle line into a median process (Pl. VI. fig. 6; text-fig. 62) instead of bearing a median posterior notch, as in *M. durobrivensis* (Pl. V. fig. 10) and *M. platyclis* (Pl. VI. fig. 3). (5) The limbs appear to be more lightly constructed than in the other species (text-fig. 63).

All the following specimens from the Leeds Collection were obtained from the Oxford Clay in the neighbourhood of Peterborough.

R. 2421 (Leeds Coll. 25). Imperfect skull and skeleton. The parts present are :—Basioccipital, exoccipitals, basisphenoid, parts of maxilla, premaxillæ, portions of parietals and frontals, portions of the mandible; forty-four cervical vertebræ, two pectorals, nineteen dorsals, and twelve sacrals and caudals; many of the vertebræ have the fused neural arches preserved, and in some of the cervicals and caudals the ribs are preserved; dorsal ribs; portions of coracoids and scapulæ; fore paddles (phalangeal portion incomplete); ilium; ischia; hind paddle (phalangeal portion incomplete). Type specimen described and figured by Seeley, *loc. cit. supra* (also figured in Plates III. & IV.).

The dimensions (in centimetres) of this specimen are :—

Skull (Pl. III. figs. 1, 2):
Basioccipital: transverse diameter of occipital condyle . . 2·2
vertical diameter of occipital condyle . . . 2·2
greatest length in middle line . . . . . . 3·9
„ width at lateral processes . . . . 5·1
Width across snout at maxillo-premaxillary suture . . . . 6·3
Mandible (Pl. III. fig. 3):
Length from anterior end of symphysis to angle . . . . 34·0
„ of symphysis . . . . 5·2
„ of post-articular region . . . . 4·1

Vertebræ: Cervicals (Pl. IV. figs. 1–4)	Atlas and axis.	Third cervical.	Fifth cervical.	Eleventh cervical.	Fifteenth cervical.	Twenty-fourth cervical.	Thirty-second cervical.	Fortieth cervical (crushed).
Length of centrum in mid-ventral line.	5·3	2·7	3·6	4·2	4·7	5·5	5·6	5·3
Posterior width of centrum .	3·3	3·5	3·7	4·3	4·9	5·4 (app.)	5·9	6·8
„ height of centrum .	3·1	3·0	3·1	3·5	3·8	4·7	5·4	5·5
Height to top of neural spine	..	..	7·8	..	..	14·4	16·9	19·8 (app.)

Vertebræ: Pectorals, Dorsals, etc.	Second pectoral.	Dorsal (forty-ninth).	Dorsal (fifty-eighth).	Dorsal (sixty-third).	Sacral (sixty-eighth).	Caudal (seventy-fifth).
Length of centrum in mid-ventral line.	4·9	5·6	5·2	4·8	3·9	3·8
Posterior width of centrum	6·6	6·0 (app.)	6·0	6·0	6·0	5·4
„ height of centrum .	5·2	5·7	4·7	4·6	4·2	4·0

Coracoid: width of each coracoid at lower angle of glenoid
    cavity . . . . 16·0 (app.)
    length from anterior angle of scapular surface to
        postero-external angle . . . . 31·5
Humerus (Pl. IV. fig. 7): length . . . . 28·3
    width of head (antero-posterior) . . 6·3
    „   „ with tuberosity . . 9·8
    „   shaft at narrowest . . . 7·0
    „   distal end . . . . . . 17·0
Radius (Pl. IV. fig. 7): greatest length . . . . 9·1
    „ width . . . . 8·3
Ulna (Pl. IV. fig. 7): greatest length . . . . 5·7
    „ width . . . . 7·4
Ischium (Pl. IV. fig. 9): length . . . . 8·1
    width from acetabulum to symphysis. 17·8
Ilium (Pl. IV. fig. 8): length . . . . 18·1
    width of upper end . . . . 5·5
    „ lower end . . . . 6·1
Femur (Pl. IV. fig. 10): length . . . . 26·3
    diameter of head . . . . 7·0
    width of upper end (with trochanter). 7·9
    „ shaft at narrowest . . . . 4·8
    „ distal end. . . . . . . 13·3
Tibia (Pl. IV. fig. 10): greatest length . . . . 6·9
    „ width . . . . 5·6
Fibula (Pl. IV. fig. 10): greatest length . . . . 5·2
    „ width . . . . 6·3

R. 2422 (Leeds Coll. 21). Greater part of a skull and skeleton of an old individual. The parts preserved are a much broken skull (basioccipital, basisphenoid, exoccipital, and opisthotic figured, text-fig. 44), a nearly complete mandible, atlas, axis, and forty-two other cervicals (in these the sutures between the centra and the neural arches and ribs are obliterated, but in most cases the ribs have been broken off), nineteen dorsals (much distorted), and twenty-four caudals (in these the arches, ribs, and, in many cases, chevrons are fused with the centra); numerous ribs, imperfect shoulder-girdle (in this ossification is complete, the coraco-scapular symphysis being continuous ; the clavicular arch is not preserved) ; fore paddles ; ilia ; ischium ; part of pubis ; hind paddles.

The dimensions (in centimetres) of this specimen are :—

Skull (text-fig. 44):
    Length from posterior border of pineal foramen to tip of
        snout . . . . 18·7
    Length of basioccipital . . . . (approx.) 4·0
    Width of basioccipital at lateral processes. . . . . . . 5·5
    Vertical diameter of occipital condyle . . . . . . 2·8
    Width of articulation of quadrate . . . . 3·0
Mandible :
    Length . . . . 34·2
    „ of symphysis . . . . (approx.) 5·4

MURÆNOSAURUS LEEDSI.

Vertebræ: Cervicals...	Atlas and axis.	Third cervical.	Fifth cervical.	Tenth cervical.	Fifteenth cervical.	Twentieth cervical.	Twenty-fourth cervical.	Posterior cervical.	Posterior cervical.
Length of centrum in mid-ventral line	6·1	3·1	4·0	4·6	5·5	6·0	6·1	5·7	5·7
Posterior width of centrum	3·6	4·0	4·0	4·8	?	5·8	5·7	7·7	8·3
Posterior height of centrum	2·7	2·9	3·2	3·7	?	5·0	5·2	6·0	6·1
Height to top of neural spine	5·8(app.)	6·9	8·1	9·7	12·2(app.)	?	?	?	17·8

The total length of the neck was about 241 cm. (about 8 ft.). The dorsal vertebræ are too much crushed for measurement, but the dimensions of an anterior and middle caudal centrum are:—Length on mid-ventral line 3·4, 3·5 ; posterior width of centrum 7·4, 6·6 ; posterior height of centrum 4·7, 4·3.

```
Shoulder-girdle : width at hinder end of glenoid cavity
                       (exaggerated by crushing) . . . . . .  43·0
                 long diameter of coraco-scapular foramen .  13·7
Humerus : greatest length . . . . . . . . . . . . .  33·2
          width of shaft (widened by crushing) . . . . .   8·3
            „    distal end    (ditto)         . . . .  19·0
Pelvis.—Ilium : length . . . . . . . . . . . . . .  19·5
                width of upper end . . . . . . . . .   5·2
                  „    lower end . . . . . . . . .    7·1
        Ischium : width of upper end (crushed) . . . . .   9·3
                  „    neck . . . . . . . . . . .    7·3
        Femur : length . . . . . . . . . . . . .   30·4
                width of shaft . . . . . . . . . .    6·3
                  „    distal expansion . . . . . . .  15·8
```

R. 2423 (Leeds Coll. 22). Incomplete skull and skeleton, the bones being much crushed and distorted. The following parts are preserved:—Basioccipital, basisphenoid, supra-occipital, part of squamosal, the articular portions of the mandible, atlas, axis, and about forty other cervical vertebræ wanting ribs and arches, nineteen dorsals and nineteen caudals, crushed neural arches, fragments of scapulæ and ischium, humeri, femora, epipodials, and other bones of paddles.

```
Length of the basioccipital . . . . . . . . . . . .  3·5
Width of occipital condyle from side to side . . . . . .  2·5
   „    basioccipital at lateral processes . . . . . . .  4·5
Length of united basioccipital and basisphenoid . . . . .  6·9
   „    axis and atlas . . . . . . . . . . . . .  5·3
```

The other vertebræ are too much distorted to give measurements of any value. The centra of the posterior cervicals appear to have been rather shorter than in the type specimen.

R. 2424 (Leeds Coll. 23). Imperfect skeleton, the skull and mandible being entirely absent. The vertebral column is represented by thirty-four cervicals, two pectorals, twenty-three dorsals (including two sacrals), and eighteen caudals. In some cases the centra are much crushed and in all are separated from the neural arches, cervical and caudal ribs, etc. Five complete neural arches (four dorsal and one caudal), with portions of several others, are preserved. Several cervical ribs and several more or less nearly complete dorsal ribs are present. The right humerus with the radius and ulna, two ischia, part of a pubis, both femora, with some other bones of the hind paddle are preserved.

In this specimen ossification is not so far advanced as in the type, although it is already larger, so that probably this individual might have attained a considerably greater size than the type. In the form of the limb-bones, so far as known, the two are closely similar, but in the cervical vertebræ, especially in the posterior members of the series, there are some differences, the most striking being that the width of the centrum is somewhat greater in proportion to the height. It must be noted, however, that in the type specimen many of the vertebræ have been distorted by pressure, and this may account for the dissimilarity.

Associated with this skeleton and in exactly similar condition of preservation is a clavicle (figured by Seeley in Proc. Roy. Soc. vol. li. (1892) p. 141, fig. 8) of an irregular triradiate form, its irregularity of outline probably indicating that the clavicles were undergoing reduction. With the exception of a portion of the opposite clavicle, this bone is the only element of the shoulder-girdle preserved in this specimen, and no such bone has been found with any other *Murænosaurus*-skeleton, so far as I am aware, so it is just possible that it may actually belong to a species of *Cryptocleidus*. If, on the other hand, this is not so, and the bone is actually that of *M. leedsi*, it indicates that in some cases the clavicles undergo much less reduction than usual, and that this species must, in the structure of its clavicular arch, have been much like *M. platyclis*.

The approximate dimensions (in centimetres) of this specimen are :—

Vertebræ: Cervicals*.........	Fifth cervical.	Tenth cervical.	Sixteenth cervical.	Twenty-sixth cervical.	Thirty-fifth cervical.	Forty-second cervical.
Length of centrum in mid-ventral line	3·4	4·2	5·1	5·2	5·3	5·2
Width at posterior end . .	3·4	4·3	4·7	5·6	6·3	6·7
Height at posterior end . .	2·6	3·4	4·1	4·8	4·8	4·8

Caudals..................................	Anterior.	Posterior.
Length of centrum on mid-ventral line	3·6	3·4
Width at posterior end	5·4	5·1
Height at posterior end	4·0	4·0

Humerus: length	27·5
width of head	6·8
greatest width of proximal end	9·0
least antero-posterior diameter of shaft	7·3
width of distal end	16·2

* The numbers of the cervicals in the series are only approximate. The dorsals are all too much crushed for measurement.

Femur: length . . . . 25·8
  width of head . . . . 6·5
  greatest width at proximal end (with trochanter) . . 7·5
  least antero-posterior diameter of shaft . . . . . 4·6
  width of distal end . . . . 12·9
Pubis: greatest width . . . . 2·9
  ,, length . . . . 2·4
  length of acetabular surface . . . . 7·5
  ,, surface for ischium . . . . 3·5
Ischium: greatest length from acetabulum to middle line
    (approx.) 18·0
  length of acetabular surface . . . . . (approx.) 4·5
  ,, iliac surface . . . . . . (approx.) 2·5
  ,, pubic surface . . . . . . (approx.) 3·5
  width at narrowest point . . . . 5·1

R. 2864 (Leeds Coll. 34). Imperfect skull and skeleton of an individual smaller than the type specimen. The portions preserved are:—Basioccipital, basisphenoid with part of parasphenoid, portions of exoccipitals, of frontals, and of premaxillæ, vomer and part of the pterygoid, quadrate; greater part of mandible; several detached teeth; forty-five cervical and pectoral vertebræ, twenty dorsal (and ? sacral) vertebræ and twenty caudals with some caudal ribs and chevrons; portions of the limb-girdles, the fore and hind paddles.

The fore and hind paddles (text-fig. 63) are the most nearly complete yet found, practically all the bones on one side having been collected and retained in their natural relative positions. The caudal vertebræ (text-fig. 59) are in excellent preservation, many of them still retaining their union with the neural arches and caudal ribs and in one case with the chevrons; of these last several are preserved separated from the centra (text-fig. 59, F, G, H).

The dimensions (in centimetres) of this specimen are:—

Skull:
  Basioccipital: transverse diameter of the occipital condyle . 2·2
    greatest length on middle line . . . . . 3·2
    ,, width at lateral processes . . . . 4·1
Mandible:
  Length from anterior end of symphysis to angle (approx.) 25·0
    ,, of symphysis . . . . 3·4
    ,, of postarticular region . . . . . (approx.) 3·6

Vertebræ: Cervicals *	Atlas and axis.	Fifth cervical.	Tenth cervical.	Fifteenth cervical.
Length of centrum	4·5	2·2	3·5	3·9
Width of hinder end of centrum .	2·5	3·1	4·0	4·4
Height of hinder end of centrum .	2·1	2·5	3·1	3·4

* The numbers of the vertebræ in the series are approximate. Behind the fifteenth the cervical vertebræ are too much distorted for measurements to be of any value; the same is the case with the dorsals.

MARINE REPTILES OF THE OXFORD CLAY.

Caudals.

Length of centrum in mid-ventral line	3·1	3·1	2·7	2·3	2·1	2·0
Width of hinder end of centrum	4·7	4·1	3·7	3·2	3·2	2·7
Height of hinder end of centrum	4·1	3·9	3·1	2·5	2·4	2·0
„ to top of neural spine	?	7·8	6·1	4·8	4·3	..
Width between outer ends of caudal ribs	14·6	13·8	..	8·2	7·0	..

Fore paddle (text-fig. 63, B): total length	72·0
Humerus: length	25·2
greatest width at upper end (with tuberosity)	9·1
least antero-posterior width of shaft	6·3
greatest width at lower end	14·1
Radius: greatest length	7·2
„ width	7·1
Ulna: greatest length	4·9
„ width	6·1

The lengths of the successive phalanges of the longest (fourth) digit are:—4·1, 4·0, 3·5, 3·3, 3·1, 2·6, 2·5, 2·0, 1·8, 1·4, 1·3, 1·1, ·8.

Hind paddle (text-fig. 63, A): total length	70·5
Femur: length	23·8
greatest width at upper end	7·4
least antero-posterior width of shaft	5·0
greatest width at lower end	12·2
Tibia: greatest length	5·0
„ width	6·5
Fibula: greatest length	4·7
„ width	5·9

The lengths of the successive phalanges in the longest (fourth) digit are:—4·1, 3·9, 4·0, 3·6, 3·4, 2·8, 2·6, 2·3, 2·0, 1·7, 1·4, 1·1, ·9.

R. 3704. Shoulder-girdle of an old individual, probably of this species. In this specimen (text-fig. 62) the ossification of the bones seems to have proceeded further than in any other; this is especially notable in the posterior region of the coracoids, which are produced backwards with well-marked postero-lateral processes and less developed median processes not seen in younger examples. The scapulæ and coracoids meet in the mid-ventral line, and the interclavicle (Pl. VI. figs. 6, 6 a) is preserved in an almost perfect state. The middle of its anterior border is marked by a shallow concave notch (a.n.), the edge of which is smooth; on either side of this the convex anterior and lateral edges of the bone are thin and with numerous small indentations. Posteriorly there seems to have been a short pointed process (p.p), the continuation backwards of a median ridge, seen on the ventral face of the bone; this ridge begins anteriorly as a broad very slightly convex surface, and narrows backwards to this posterior point. The ventral surface seems to have been gently convex from side to side, the dorsal surface concave in the same direction. Crossing the dorsal face of the bone from side to side, at about two-thirds of its length from the anterior end, is a broad convexity, behind which the bone is slightly concave from before backwards.

The dimensions (in centimetres) of this shoulder-girdle are :—

Total length of the whole on middle line	75·0
Scapula : greatest length	33·6
length from anterior angle to posterior end of ventral ramus	24·6
length on a straight line from posterior angle of ventral ramus to tip of dorsal ramus	26·6
approx. width of glenoid surface from before backwards	5·2
approx. width of glenoid surface from above downwards	4·6
Coracoids: greatest length	48·9
width of united bones at hinder angle of glenoid cavity	40·8
width of united bones at narrowest	28·2
,, ,, between the posterior external angles (approx.)	4·9
antero-posterior diameter of coraco-scapular foramen	13·2
transverse diameter of ditto (approx.)	9·4
Interclavicle (Pl. VI. fig. 6): length (approx.)	11·9
width (approx.)	14·2

R. 2443 a. Left femur probably of this species, figured in Phillips, 'Geology of Oxford, etc.,' p. 317, text-fig. cxxi., as belonging to *Pliosaurus grandis*.

Murænosaurus durobrivensis, Lydekker, sp.

[Plate V.; text-figs. 43, 45, 49-57, 60, 65, 67.]

1889. *Cimoliosaurus durobrivensis*, Lydekker, Catal. Foss. Rept. Brit. Mus. pt. ii. p. viii.
1895. *Murænosaurus plicatus*, Andrews, Ann. Mag. Nat. Hist. [6] vol. xvi. p. 429 (pars).

Type Specimen.—An imperfect skeleton, including thirty cervical vertebræ, fifteen dorsals, and some caudals, wanting ribs and arches; shoulder-girdle wanting the clavicular arch ; imperfect fore paddle ; pelvis including both ilia and ischia and the pubis; imperfect hind paddles (R. 2428, Leeds Coll. 28). The pectoral and pelvic girdles have been described and figured in Ann. Mag. Nat. Hist. [6] vol. xvi. (1895) p. 429 (see also Pl. V.; text-figs. 43, 45, 49-57, 60, 65, 67).

This species was established by Mr. Lydekker for the reception of an Oxford Clay Plesiosaur which, while closely resembling his *Cimoliosaurus plicatus*, differs in wanting the median bony bar uniting the coracoids and scapulæ, and in possessing cervical vertebræ with relatively shorter centra, especially in the posterior portion of the neck. The absence of the median bony bar in the shoulder-girdle is probably merely an age-character, but the shortness of the cervical vertebræ is a well-defined peculiarity and is accompanied by others which distinguish this species from the other members of

128   MARINE REPTILES OF THE OXFORD CLAY.

the genus. One of these peculiarities is, that ossification does not become complete (as indicated by the fusion of the cervical and caudal ribs and of the neural arches,

Text-fig. 67.

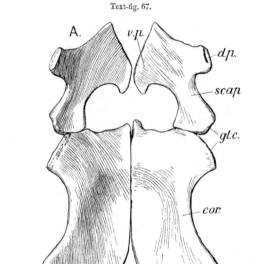

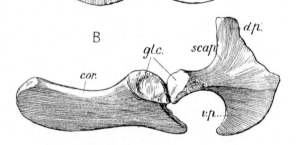

Shoulder-girdle of *Murænosaurus durobrivensis*: A, from above; B, from right side. (Type specimen, R. 2428, ½ nat. size.)

cor., coracoid; *d.p.*, dorsal ramus of scapula; *gl.c.*, glenoid cavity; *scap.*, scapula; *v.p.*, ventral ramus of scapula.

and the condition of the proximal ends of the humerus and femur) until a much greater size has been attained than in *M. leedsi*, and all parts of the skeleton are more massive than in that form. Another point is, that the expansion of the distal end of the humerus is greater than in *M. leedsi*.

As already noticed, the cervical vertebræ are shorter than in *M. leedsi*; and this difference is most marked in the posterior part of the neck, where the centra in the present species are shorter in comparison with those of the anterior region. The plications on the edges of the centra, just outside the articular faces, are particularly well marked in this species, certainly much more so than in *M. platyclis*, where they seem to be replaced by irregular rugosities. It was the existence of these plications on a cervical vertebra that caused Phillips to name one of the Oxford Clay Plesiosaurs *Plesiosaurus plicatus*: this no doubt is a *Murænosaurus*, but it is not possible on the available evidence to be sure whether it is identical with the present species or with *M. leedsi* or is a distinct form. In the pectoral girdle in this species the median union of the scapulæ with one another does not take place till a very large size has been attained, and the median junction of the coracoids and scapulæ was still later. The interclavicle is a small oval bone, with anterior and posterior rounded notches in the middle line; it is thickened in the middle, especially towards the anterior border, but laterally it thins out to a sharp edge; no trace of clavicles has been observed.

R. 2428 (Leeds Coll. 28). Portion of a skeleton, including twenty-nine cervicals (Pl. V. figs. 1–3), two pectorals, fourteen dorsals (Pl. V. figs. 4, 5), and some caudals (Pl. V. figs. 6, 7); coracoids and scapulæ (text-fig. 67); imperfect fore paddles; ilia, ischia, and one pubis (text-fig. 65); incomplete hind paddles. Type specimen referred to by Lydekker in Catal. Foss. Rept. Brit. Mus. pt. ii. (1889) p. viii. The pectoral and pelvic girdles have been described and figured in Ann. Mag. Nat. Hist. [6] vol. xvi. (1895) as those of of *M. plicatus*, Phillips, sp.

In this specimen the neural arches are wanting in all but two or three of the vertebræ, and in the cervical region the ribs are also missing, fusion with the centrum not having taken place in either case. The ventral rami of the scapulæ have not yet reached the middle line and are widely separated from the coracoids, so that there is no doubt that although this was a large individual it had not nearly attained its full size.

The dimensions (in centimetres) of some of the bones of this skeleton are as follow :—

Vertebral centra. (Pl. V. figs. 1–7.)	Anterior cervicals.		Middle cervicals.		Posterior cervicals.		First dorsal crushed.	Dorsals.		Caudals anterior.	Caudals posterior.	
Length in mid-ventral line	2·7	4·0	4·9	5·0	5·3	5·2	5·5	5·5	5·7	4·2	2·8	2·5
Width of posterior articular surface	3·1	4·2	5·7	6·0	7·1	7·2	7·4	7·2	7·2	6·5	4·6	3·7
Height of posterior articular surface	2·5	3·4	4·6	5·2	5·7	5·7	6·2	6·0	6·4	4·9	3·5	3·1

Scapula (text-fig. 67): greatest length . . . . 27·4
length from anterior angle to posterior
end of ventral ramus . . . . . 16·8

Scapula (text-fig. 67): length in straight line from posterior
 angle of ventral ramus to summit of
 dorsal ramus . . . . 26·4
 width of glenoid surface from above
 downwards . . . . 4·8
 width of glenoid surface from before
 backwards . . . . 5·4
Coracoids (text-fig. 67): greatest antero-posterior length . . 39·2
 width of united bones at posterior
 angle of glenoid cavity . . . . 38·5
 width of united bones at narrowest . 26·9
 ,, between postero-external angles 36·8
Greatest length of united coracoids and scapulæ . . . . . 60·0
Humerus (Pl. V. figs. 11, 11 a, 11 b): length . . . . 31·5
 greatest width of upper end 10·6
 width of shaft at narrowest 7·8
 ,, distal end . . . 19·4
Pelvis (Pl. V. figs. 8, 8 a, 8 b, 9, 9 a; text-fig. 65).
 Ilium : length . . . . 18·7
 greatest width of upper end . . . . 5·6
 width of shaft . . . . 3·2
 greatest width of lower end . . . . 6·2
 Pubis : greatest length . . . . 25·0
 ,, width . . . . 27·8
 length of acetabular surface . . . . 8·0
 ,, symphysial border . . . . 16·2
 Ischium : width of upper end . . . . 9·0
 ,, neck . . . . 6·5
 ,, lower expansion . . . . 20·2
 length of acetabular surface . . . . 3·8
Femur (Pl. V. figs. 12, 12 a, 12 b): length . . . . 28·4
 width of upper end . . . 8·0
 ,, middle shaft (at
 narrowest) . . 5·4
 ,, lower end . . . 15·8
Tibia (Pl. V. figs. 12, 12 a): length . . . . 7·1
 width . . . . 9·0
Fibula (Pl. V. figs. 12, 12 a): length . . . . 6·5
 width . . . . 7·2

R. 2863 (Leeds Coll. 29). Imperfect skeleton, including fragments of skull and mandible ; forty-two cervical vertebræ (text-figs. 49–51), two pectorals (text-fig. 52), twenty-one dorsals (text-figs. 52–55), and eight sacrals (text-fig. 56) and caudals (text-fig. 57) ; numerous ribs (cervical, dorsal (text-fig. 60), and caudal), ventral ribs (text-fig. 60); coracoids, scapulæ and interclavicle (Pl. V. fig. 10), humerus and some other bones of fore paddle ; ilia, ischia, pubes ; femora and some other bones of hind paddle.

In this specimen the neural arches with the neural spines are preserved united with the centra in the cervical and dorsal regions, although the neuro-central sutures are still open; in the sacral and caudal regions the arches are lost in all cases but one (text-fig. 57). In the pectoral girdle the scapulæ met in median symphysis, but did not yet extend back in the middle line to meet the coracoids, the interval between them being, however, only about 4 cm. The heads of the humeri and femora are showing signs of becoming rounded owing to the extension of ossification into the originally cartilaginous ends : in the type specimen they are nearly flat. The whole skeleton shows very well the massiveness characteristic of this species.

The dimensions (in centimetres) of this specimen are :—

Skull too fragmentary to measure.
Mandible : width of articular surface . . . . 2·6
length of postarticular region . . . . 3·8

Vertebræ: Cervicals	Atlas and axis.	Third.	Fifth.	Eleventh.	Twentieth.	Thirty-fifth.	42nd, 43rd (last true cervical).
Length of centrum in mid-ventral line	5·0	3·1	3·3	4·3	5·2	5·3	5·1
Width of posterior face of centrum	3·0	3·1	3·5	4·2	5·2	6·6	7·7
Height of posterior face of centrum	2·6	2·7	3·0	3·7	4·8(app.)	5·7	6·0
Height to top of neural arch	..	..	..	10·0	15·3	18·0	..

Pectorals and dorsals, sacrals and caudals	Second pectoral (figured).	First dorsal.	Anterior dorsal (figured).	Anterior dorsal.	Post. dorsal.	Post. (? last) dorsal.	Sacral (figured).	Anterior caudal.	Middle caudal (figured).
Length of centrum in mid-ventral line	5·6	5·7	5·6	5·9	5·6	5·4	4·7	3·4	2·9
Width of posterior face of centrum	7·8	7·6	6·9	7·6	6·9	8·5	6·7	6·3	5·4
Height of posterior face of centrum	6·3	6·2	5·9	6·5	6·0	5·8	5·1	4·7	4·1
Height to top of neural arch	..	..	20·0	22·0	20·0	..	..	..	8·3
Width between ends of transverse processes	..	..	14·5	18·4	14·6	10·6(app.)	..	..	.

Coracoid : greatest length . . . . 41·0
width of united bones at hinder angle of glenoid cavity . . . . 40·4
width at narrowest of the two united bones (approx.) 28·4
„ between angles of postero-external processes . 34·0
Scapula : greatest length . . . . 27·5
length from anterior angle to posterior end of ventral ramus . . . . 18·2
length from end of median coracoid process to tip of dorsal ramus . . . . 27·5
width of glenoid surface (above downwards) . . . 6·1
„   „   „ (before backwards) . . . 6·5
Interclavicle (Pl. V. fig. 10) : greatest length . . . . . 8·5
length on mid-line . . . . . 6·6
width (so far as preserved) . . 9·8

MARINE REPTILES OF THE OXFORD CLAY.

Humerus: length . . . . 31·0
          greatest width at upper end . . . . 11·5
          diameter of head. . . . . (approx.) 8·4
          width of shaft at narrowest . . . . 8·3
          ,,   dorsal end . . . . 19·1
Ilium: length . . . . 20·0
          greatest width at upper end . . . . 6·4
          width of shaft . . . . 3·5
          greatest width of lower end . . . . 6·7
Pubis: greatest length . . . . 26·5
    ,,   width . . . . 30·5
    length of acetabular surface . . . . 7·3
    ,,   symphysial border . . . . . (approx.) 20·0
Ischium: width of upper end . . . . 9·3
       ,,   shaft . . . . 6·1
       ,,   lower expansion . . . . 21·2
    length of acetabular surface . . . . (approx.) 5·5

R.2861 (Leeds Coll. 18). Imperfect skeleton of a large individual probably of this species, though the plications round the edges of the cervical centra are less marked than usual. The parts preserved are an imperfect and much broken skull (basioccipital, exoccipital, opisthotic, and supraoccipital, figured in text-figs. 43, 45); a nearly complete mandible; atlas, axis, and twenty-seven other cervicals, mostly with the fused arches and ribs broken away; nine dorsal and two caudal vertebræ; about twenty dorsal and cervical ribs, some portions of ventral ribs; portions of the acetabular region of the pelvis with the constituent bones united; left femur, both tibiæ and fibulæ, tarsals, and sixty-three metapodial bones and phalanges.

Some dimensions (in centimetres) of this specimen are:—

Skull (text-figs. 43, 45):
    Length in mid-dorsal line from posterior end of the parietals
        to the tip of the snout . . . . 26·4
    Length from pineal foramen to tip of snout . . . . . . 20·9
    Transverse diameter of occipital condyle . . . . 2·9
    Vertical diameter of occipital condyle . . . . 2·9
    Length of basioccipital . . . . 4·3
    Greatest width at lateral processes . . . . 5·6
Mandible: length . . . . 37·9
      ,, of symphysis . . . . (approx.) 5·0
      width of articular surface for quadrate . . . . 3·1

Vertebræ	Atlas and axis.	Fifth cervical.	Tenth cervical.	Twentieth cervical.	Twenty-fifth cervical.	Twenty-ninth cervical.	Mid-dorsal.	Mid-dorsal.	Anterior caudal.	Middle caudal.
Length of centrum in mid-ventral line . .	6·0	4·8	5·0	6·0	6·3	6·3	6·3	6·2	4·9	3·2
Width of posterior face .	4·0 app. 4·8		4·9	6·5	7·1	7·5	8·1	7·9	8·3	5·9
Height of posterior face .	3·1	4·1	4·2	5·3	5·8	6·3	7·2	6·9	5·4	4·9

Femur: length . . . . 35·7
diameter of head (crushed) . . . . 7·5
least antero-posterior diameter of shaft . . . . . 6·7
width of distal end . . . . 16·4
Tibia: width . . . . 9·2
length . . . . 7·3
Fibula: width . . . . 8·4
length . . . . (approx.) 6·7

R. 2427 (Leeds Coll. 27). Portions of the skeleton of a rather small individual, probably of this species. The ossification is complete, although the animal is smaller than the type specimen (R. 2428) or than R. 2863. The parts preserved are basioccipital and basisphenoid, exoccipitals, premaxilla; parts of mandible including the symphysial region; the atlas and axis, and thirty-six other cervical vertebræ, the fused neural arches and cervical ribs being broken away in nearly all cases; eleven dorsal and thirteen caudal vertebræ, some with the arches and ribs; numerous ribs, dorsal and ventral, chevrons; imperfect shoulder-girdle, ischium, (?) femur, tibia, and fibula.

As already mentioned, although the ossification in this specimen is far advanced, the neural arches, the cervical and caudal ribs being fused with the centra, and the coracoscapular foramen completely closed by the median junction of the coracoids and the ventral bar of the scapulæ, nevertheless it is smaller than the type specimen, in which none of these indications of maturity are present: it is possible that this difference in size may be a sexual one, but the evidence is not sufficient to make this certain. In this specimen the caudal vertebræ and chevrons are well preserved: these show that on the anterior caudal region the chevron-facets are single and confined to the posterior border of the centrum, where they may be raised on a well-marked prominence. Further back the chevrons bear two facets, which in some cases may be almost completely separated from one another; in this case they articulate between the centra of two successive vertebræ, which bear corresponding oblique facets both on their anterior and posterior borders. The chevrons of opposite sides did not unite ventrally.

The dimensions (in centimetres) of this specimen are:—

Basioccipital: transverse diameter of occipital condyle . . . 2·5
vertical diameter of occipital condyle . . . . 2·6
greatest length in middle line . . . . . . 4·1
„ width at lateral processes . . . . . 5·2
Mandible: length of symphysis . . . . 4·0
„ postarticular process . . . . 3·7
width of surface for articulation with quadrate . 2·5

Vertebræ......	Atlas and axis.	First cervical.	Tenth cervical.	Twentieth cervical.	Thirtieth cervical.	Anterior dorsal.	Middle dorsal.)	Sacral.	Anterior caudals.	Middle caudal.
Length of centrum in mid-ventral line .	5·5	2·8	4·2	5·2	5·3	5·3	5·7	5·0	4·2 3·8	3·0
Width of posterior face of centrum .	3·3(crushed)	3·6	4·6	6·1	6·6	6·3	6·7app. 6·1		5·4 5·3	4·2
Height of posterior face of centrum .	. .	. .	3·7	4·8	5·2	5·6	6·2	5·0	4·5 4·5	3·6

The numbers of the cervicals in the series are only approximate.

Shoulder-girdle: width of coracoids between posterior angles
of the glenoid cavities . . . . 42·0
Ischium: width of upper end . . . . 11·3
" neck at narrowest . . . . 6·7

Murænosaurus platyclis, Seeley.

[Plate VI. figs. 1-5; text-figs. 48, 58, 68, & 69.]

1892. *Murænosaurus platyclis*, Seeley, Proc. Roy. Soc. vol. li. pp. 139-141, figs. 6, 7.

Type Specimen.—A nearly complete skeleton, including skull (Pl. VI. fig. 1), mandible (Pl. VI. fig. 2), thirty-eight cervical vertebræ (Pl. VI. figs. 4, 5), twenty-one dorsals, fourteen caudals, numerous dorsal and ventral ribs, shoulder-girdle (coracoids incomplete posteriorly), humeri and other bones of the fore paddles, pelvic girdle, femora. The shoulder-girdle (Pl. VI. fig. 3; text-fig. 68) is described and figured by Seeley, *loc. cit. supra*.

Text-fig. 68.

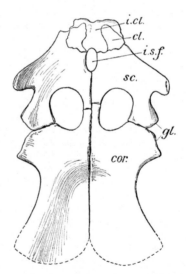

Shoulder-girdle of *Murænosaurus platyclis*, from above.
(Type specimen, R. 2678, ⅛ nat. size.)

l., clavicle; *cor.*, coracoid; *gl.*, glenoid cavity; *i.cl.*, interclavicle; *i.s.f.*, interscapular foramen; *sc.*, scapula.

This species was established by Professor Seeley on the evidence of the shoulder-girdle only, but its skeleton differs in several other respects from those of *M. leedsi* and *M. durobrivensis*. From the former it, like *M. durobrivensis*, differs in having attained a much greater size and more massive proportions, and in possessing shorter centra in the cervical vertebræ, which, however, in the hinder region of the neck are not quite so much shortened as in *M. durobrivensis*. In the anterior cervicals at least

Text-fig. 69.

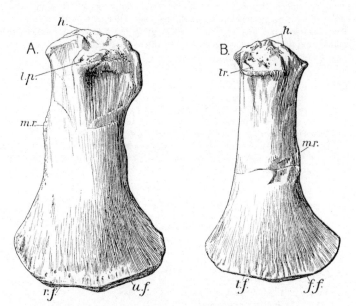

Left humerus and left femur of *Murænosaurus platyclis*: A, humerus; B, femur.
(Type specimen, R. 2678, ⅓ nat. size.)

f.f., facet for fibula; *h.*, head; *l.p.*, tuberosity; *m.r.*, ridge for muscle-attachment; *r.f.*, facet for radius; *t.f.*, facet for tibia; *tr.*, trochanter; *u.f.*, facet for ulna.

the centra are a little broader in proportion to their height than in *M. durobrivensis*. As in *M. leedsi*, the longitudinal ridge on the side of the cervical centrum, above the articular facet for the rib, is much more strongly developed than in *M. durobrivensis*, in which it may be completely wanting. Another point distinguishing the cervical vertebræ from those of the other species is the great height and width of the neural spines on the hinder part of the neck. The shoulder-girdle (Pl. VI. fig. 3; text-fig. 68) is ossified completely on the Elasmosaurian plan. The scapulæ are notable for the great antero-posterior extent of the ventral rami ; these unite in a long symphysis posteriorly, but are separated anteriorly by a deep notch, which was covered in front by the overlying interclavicle, while the posterior part of the notch forms the greater part of the interscapular foramen ($i.s.f$), closed in front by the posterior border of the interclavicle. The clavicles are represented by thin plates of bone adherent to the visceral face of the interclavicle ; their free portions are not preserved in any specimen. The humerus and femur seem to be proportionately rather shorter and stouter than in *M. leedsi*.

The skeleton in this species may attain a great size and degree of massiveness, equalling, or perhaps even surpassing, in this respect the largest specimens of *M. durobrivensis* (see measurements of R. 2425).

R. 2678 (Leeds Coll. 12). Greater part of a skeleton of a large individual in which the ossification of the shoulder-girdle and the fusion of the cervical ribs are complete. The parts preserved are:—Skull (Pl. VI. fig. 1), mandible (Pl. VI. fig. 2; text-fig. 48), thirty-eight cervical vertebræ (Pl. VI. figs. 4, 5), twenty-one dorsals, fourteen caudals, numerous dorsal and caudal ribs, (?) chevrons, shoulder-girdle (Pl. VI. fig. 3 ; text-fig. 68), humeri (text-fig. 69 A), radius, ulna, and other bones of the paddle, pelvic girdle, femora (text-fig. 69 B). Type specimen described and figured by Seeley in Proc. Roy. Soc. vol. li. (1892) pp. 139–141, figs. 6, 7.

The dimensions (in centimetres) of this specimen are :—

Skull (Pl. VI. fig. 1):
Approximate length on mid-dorsal line . . . . 27·0
Length from pineal foramen to tip of snout . . . . . . 19·3
Width of frontals between the orbits . . . . (approx.) 4·6
Vertical diameter of the occipital condyle . . . . . . 2·8
Length of basioccipital . . . . 4·4
„ basisphenoid . . . . 3·8
Mandible (Pl. VI. fig. 2 ; text-fig. 48) :
Extreme length . . . . (approx.) 41·0
Length of symphysis (crushed) . . . . 6·0
Depth at coronoid process . . . . 7·4
Width of articular surface for quadrate . . . . 3·0

Cervical vertebræ (Pl. VI. figs. 4, 5, 5 a, 5 b)	Anterior. (Fig. 5.)		Middle.		Posterior. (Fig. 4.)	
Length of centrum in mid-ventral line	4·4	4·8	5·0	5·8	6·2	6·3
Width of posterior end of centrum	4·9	5·3	5·5	6·4	?	?
Height of ditto	3·9	4·2	4·4	5·3	?	?
Height to top of neural spine	9·3	9·6	10·2 (app.)	..	18·8	21·0

The dorsal vertebræ are too much crushed for useful measurements to be made.

Anterior caudal vertebræ:
Length in mid-ventral line . . . . 4·4   4·2
Width of posterior end of centrum . . . . . 6·2   6·4
Width to end of rib-facets . . . . 7·8   7·9
Height of posterior end of centrum . . . . . 5·3   5·1
Height to top of arch . . . . 7·8   ?

Shoulder-girdle (Pl. VI. fig. 3; text-fig. 68):
Width at hinder angle of glenoid cavity (exaggerated by crushing) . . . . 50·0
Greatest length of interclavicle . . . . 11·0
Length on middle line of interclavicle . . . . 7·3
Greatest width of interclavicle . . . . 20·5
Greatest length of scapula . . . . 35·7
Length of scapular symphysis . . . . 17·0
Antero-posterior diameter of coraco-scapular foramen . . 15·0
Transverse diameter of ditto . . . . 11·5

Humerus (text-fig. 69 A):
Length . . . . 33·7
Greatest width at upper end (crushed) . . . . 12·7
Least antero-posterior diameter of shaft . . . . 9·1
Greatest width at lower end . . . . 20·4

Radius: length . . . . 9·9
         width at upper end . . . . 10·0

Ulna: length . . . . 7·5
       width . . . . 8·5

Ilium: length . . . . 20·6
        width at lower end (crushed) . . . . 7·9

Femur (text-fig. 69 B): length . . . . 31·4
                        width at upper end . . . . . . 8·3
                        least antero-posterior diameter of shaft . 6·8
                        width of distal end . . . . . . 18·5

R. 2425 (Leeds Coll. 24). Portions of a skeleton of a very large individual with the bones, which are uncrushed, of exceptionally massive structure. The portions preserved are:—the posterior part of the right ramus of the mandible, a tooth, nine cervical vertebræ (the anterior ones with the longitudinal lateral ridge well developed and the centra a little longer than in *M. durobrivensis*), two dorsals, five caudals (text-fig. 58), numerous dorsal (about 34) and ventral (about 18) ribs, ? chevrons, humeri, right ilium, femora, tibia and fibula, and numerous bones of the paddles.

The dimensions (in centimetres) of this specimen are:—

Length of the mandible from posterior angle to coronoid
process of articular-surangular . . . . 14·5
Width of articular surface for quadrate . . . . 3·0

Vertebræ	Anterior cervical.	Middle cervical.	Posterior cervical.	Middle dorsal.	Anterior caudal.	Posterior caudals. (Figured.) (Figured.)		
Length of centrum in mid-ventral line	4·5	6·5	5·9	6·8	5·0	3·5	2·9
Width of posterior face of centrum	5·2(app.)	7·7	8·7	8·4	7·0	4·0(app.)	3·6
Height of ditto	..	6·0	6·7	7·6	5·9	5·2	2·8

Humerus: greatest length . . . . 36·7
         longest diameter of head . . . . 11·0
         greatest width of upper end . . . . 14·3
         antero-posterior diameter of shaft at narrowest . 9·7
         width of distal end . . . . 23·7
Ilium: greatest length . . . . 21·3
      width of upper end . . . . 7·3
      ,, middle of shaft . . . . . . (approx.) 4·0
      ,, lower end . . . . 7·8
Femur: greatest length . . . . 34·9
      long diameter of head . . . . (approx.) 9·8
      greatest width of upper end . . . . . (approx.) 12·2
      width of shaft at narrowest . . . . 7·3
      width of distal end . . . . 19·6

R. 2456. Portions of a skeleton of a fully adult individual probably of this species. The cervical vertebræ are greatly crushed and distorted, so that their characters cannot be made out with certainty. The parts preserved are :—ten cervical vertebræ, eleven dorsals and fourteen caudals, mostly wanting the neural arches and much crushed and distorted ; portions of the coracoids and scapulæ, an imperfect interclavicle with traces of the thin adherent clavicle on one side, humeri, radii, ulnæ, and other bones of the fore paddle ; ilium, left femur, tibia and fibula, and other bones of the hind paddle.

In this specimen ossification is very far advanced : the coracoids and scapulæ meet in a continuous median symphysis, and the humerus and femur are notable for the great development of the rugosities for the insertion of muscles on the ventral face of their shafts.

The dimensions (in centimetres) of this specimen are :—

Humerus: length . . . . 33·6
         greatest width of upper end (crushed) . . . . . 11·0
         width of shaft at narrowest . . . . 8·4
         width of distal end . . . . 19·7
Radius: greatest length (anterior border) . . . . 9·0
       width at proximal articular surface . . . . . . 9·5
Ulna: greatest length . . . . 7·1
    width at proximal articular surface . . . . 8·0
Ilium: length . . . . 22·5
     ,, lower end (crushed) . . . . 6·4

Femur: length	32·0
greatest width of upper end (crushed)	9·7
width of shaft at narrowest	6·0
„ distal end	16·2
Tibia: greatest length	6·7
„ width	8·0
Fibula: greatest length	6·6
„ width	8·5

The vertebræ and shoulder-girdle are too much crushed and imperfect to give any measurements of value.

Genus PICROCLEIDUS, Andrews.

[Ann. Mag. Nat. Hist. [8] vol. iv. (1909) p. 421.]

Plesiosaurs in which ossification is complete while they are still of very small size. Skull known only from a few fragments, resembling generally the same parts of the skull of *Tricleidus seeleyi*. The neck includes upwards of thirty-nine vertebræ, the centra of which are shorter than in *Murænosaurus* (especially in the anterior region) and longer than in *Cryptocleidus*. The ends of the centra are considerably wider than high and are almost flat, often with a small mammilla in the centre. The single-headed cervical ribs of the anterior part of the neck have a distinct anterior limb (Pl. VII. fig. 5, *a.p.*), which further back in the series may be reduced to a small angular projection on the anterior border. The neural spines on the anterior part of the neck are low and rather wide, but they increase gradually in height till in the hinder region they are both wide and high (Pl. VII. fig. 3). In the shoulder-girdle (Pl. VII. figs. 2, 2*a*; text-fig. 70) the clavicular arch consists of a small interclavicle shaped somewhat like an arrow-head and triangular in section; the clavicles, if present at all, seem to be represented by mere films of bone, adherent to the visceral face of the scapulæ. These are of a typically Elasmosaurian type, meeting in the middle line in an extensive suture and extending back to meet the anterior median prolongation of the coracoids. Anteriorly they are separated by a notch for the reception of the interclavicle. The coracoids are comparatively short, posteriorly they are produced backwards into blunt processes, both at their outer and inner angles. The humerus is only slightly expanded distally and articulates with the radius and ulna only. These bones show a tendency to greater elongation than is usual in the family. The pelvis is imperfectly known; it seems to resemble that of *Murænosaurus* rather closely, but the expanded blade of the ischium is relatively longer.

This genus is distinguished from *Cryptocleidus* and *Tricleidus* by the greater number of cervical vertebræ in the neck and the flatness of their terminal faces; the shoulder-girdle also differs widely from the types characteristic of those genera. From

Text-fig. 70.

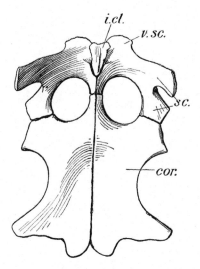

Shoulder-girdle of *Picrocleidus beloclis*, from above, rather less than ¼ nat. size.
(Type specimen, R. 1965.)

cor., coracoid; *i.cl.*, interclavicle; *sc.*, scapula; *v.sc.*, ventral ramus of scapula.

Murænosaurus it is distinguished by the shortness of the centra of the cervicals and the form of the shoulder-girdle.

Two species are known at present.

Picrocleidus beloclis, Seeley, sp.

[Plate VII.; text-fig. 70.]

1892. *Murænosaurus beloclis*, Seeley, Proc. Roy. Soc. vol. li. pp. 143–145, figs. 10–12.
1909. *Picrocleidus beloclis*, Andrews, Ann. Mag. Nat. Hist. [8] vol. iv. pp. 421–423, fig. 3.

Type Specimen.—A portion of a skeleton including six cervical vertebræ, two dorsals, ten ribs, scapulæ, coracoids, interclavicle, humeri, one radius, ulnæ (R. 1965, Leeds

Coll. 14). Shoulder-girdle and radius and ulna figured by Seeley, *loc. cit. supra*, figs. 10-12.

This species was distinguished by Professor Seeley on the strength of the structure of its clavicular arch, and its numerous other peculiarities show that he was fully justified in his conclusions.

In the type specimen no parts of the skull are preserved, but in the case of a second individual (R. 3698, Leeds Coll. 173) referred to this species, some fragments of the skull and mandible are present, the chief elements being the basioccipital (still articulating with the atlas and axis), the quadrates with part of the squamosals attached, and a much crushed exoccipital. The *basioccipital* is much distorted, but, apart from the small size, does not seem to differ in any important particular from the same bone in *Tricleidus* (p. 150). The *quadrate* is notable for the narrowness of the articulation for the mandible, in this respect differing widely from the same part in *Tricleidus* and *Murænosaurus*, the outer convex condyles seen in the quadrates in those genera being almost absent: this peculiarity is, of course, reflected in the articular surface of the mandible, the outer concavity being correspondingly narrow. The postarticular (angular) region of the mandible is much more slender than in the other genera. A few fragments of the crowns of teeth on the back of the right ramus of the mandible show that the enamel is raised into fine longitudinal ridges.

In the type specimen only six cervicals and two dorsals are preserved, but in R. 3698 all the cervicals (Pl. VII. fig. 5), with the possible exception of one or two posterior ones, and some sixteen caudal vertebræ, are preserved. The cervicals, including the atlas and axis, are at least thirty-nine in number. The *atlas* and *axis* (Pl. VII. fig. 5) are closely fused together and are relatively much shorter than in *Murænosaurus* and *Tricleidus*, although their structure seems to be the same as in those genera. Both atlas and axis are rib-bearing ($r\,1, r\,2$); the neural arch and spine are lower than in the above-mentioned genera, and there is no hypapophysial ridge. The posterior face of the axis is gently concave and is much wider than high, characters that are also present in the centra of the other cervical vertebræ. In these the centra are considerably wider than they are long, and their upper surface is deeply grooved by the floor of the neural canal (figs. 4, 5 a, 5 b). The neural spines are low and inclined backwards in the anterior region, but further back they become more upright and at the same time broad and high (Pl. VII. figs. 3, 4, 5). The anterior cervical ribs have a well-marked anterior process (*a.p.*), which further back is reduced to a slight angulation on the front of the rib. Two anterior dorsals are preserved in the type specimen: in these the centra are wide and depressed, the ends being very slightly concave with a gently raised central area. The transverse processes are very stout and strong and terminate in a broad facet for the single-headed dorsal rib; the neural arches are not well preserved. Some of the caudal vertebræ are shown in the second specimen: of these the anterior ones have wide low centra (Pl. VII. figs. 6, 6 a) with somewhat concave ends; on the sides they bear, on slight prominences, large rounded facets for the caudal ribs; the ventral face is very gently convex from side to side, and may or may not be perforated by nutritive foramina. Further back the caudal centra become deeper and narrower, the greater depth being mainly due to the development of a pair of strong longitudinal ventral ridges, the posterior ends of which are obliquely truncated by facets for the chevrons. Between these ridges the surface of the centrum is concave from side to side, as it is also between the ridges and the costal facets. The ends of the centra in this region are more strongly concave than further forwards. In the hinder part of the tail the ventral ridges become very strongly marked, and towards the end of the column they are produced into strong tuberosities which may coexist with chevron-facets, and therefore are not fused chevrons; these last-mentioned elements extend to the termination of the tail. In the caudal region the neural arches are low with stout pedicles, which unite with the anterior three-fourths of the centrum; the neural spines are short, stout, and inclined backwards. In the middle of the tail the zygapophyses are well developed, but further back they are reduced and at last disappear, and in a few of the terminal caudal vertebræ it is doubtful whether the neural arch was present at all.

A number of small bones of peculiar form, associated with the caudal vertebræ, seem to be caudal ribs. If this determination be correct, it would appear that the ribs of some of the caudal vertebræ united with one another at their outer ends by facets and in some cases even fused; it might be suggested that these bones were chevrons, but their articulations do not seem to fit the chevron-facets.

The *shoulder-girdle* (Pl. VII. figs. 2, 2 *a*; text-fig. 70) in the type specimen is well preserved. The clavicular arch has been figured and described in detail by Professor Seeley [*], who states that "The interclavicle was found *in situ*, resting on the visceral surface in a depression between the anterior margins of the scapulæ and not projecting in advance of those bones. It is lanceolate in contour, $2\frac{3}{4}$ inches long, $1\frac{3}{4}$ inches wide towards the slightly concave anterior margin, and half as wide at the rounded posterior extremity. It is a little distorted, like the other bones of the shoulder-girdle, has a flat visceral surface and an angular ventral surface, due to the bone being traversed by an elevated median ridge, which dies away anteriorly, and from this ridge the lateral surfaces are inclined. On the left side of the ventral surface its middle part is covered by a thin film of bone, which I suppose may be part of the clavicle. It corresponds in texture and thickness with a detached film of bone which rests upon the right scapula. That ossification is triangular, about $1\frac{1}{4}$ inch in each measurement, and has nearly straight sides. It is quite separate from the interclavicle and lies towards the external border of the scapula; there is no surface for its articulation, for all the

---

[*] Proc. Royal Soc. vol. 51 (1892) p. 143.

margins of the interclavicle are sharp, thin, and perfectly ossified, like its median crest. It is therefore probable that the clavicles were either loosely articulated to its margin, or extended between the interclavicle and scapula."

The *interclavicle* (*i.cl.*) thus described is shown on Pl. VII. figs. 2, 2 *a*. In section it is nearly triangular, and the ventral angle no doubt fitted between the divergent anterior ends of the ventral rami of the scapulæ; possibly, however, these were already united by cartilage, so that the interclavicle would be completely shut in below. The film-like patches of bone are probably remnants of the disappearing clavicles, as Seeley suggests: in *Murænosaurus platyclis* (see Pl. VI. fig. 3) a somewhat similar condition is found, though the reduction has not been carried nearly so far.

The *scapulæ* (*sc.*, Pl. VII. fig. 2) are similar to those of *Tricleidus*, except that the ventral rami (*v.sc.*) are relatively smaller. Their sutural union in the middle line is wide and strong, and posteriorly is continuous with the median suture of the coracoids; anteriorly it is limited by a deep notch, and in front of this is the cleft occupied by the interclavicle. The dorsal ramus (*d.sc.*) is large, but neither it nor the posterior part of the bone presents any important peculiarities.

The *coracoids* (*cor.*, Pl. VII. fig. 2) are smaller, and especially shorter, in proportion to the girdle as a whole than in the other genera. As usual in the group, they unite in a long median suture which extends forwards to the scapulæ, completely dividing the coraco-scapular openings. Between the glenoid cavities (*gl.*) the bones are much thickened, particularly towards the middle line, the symphysis here being very deep. Posteriorly the bones are thin. The surface for union with the head of the scapula is triangular and makes an angle of 120° with the glenoid surface, which with the glenoid surface of the scapula makes a deep oval fossa for the articulation of the humerus. Behind the glenoid cavity the lateral border of the coracoid is deeply concave. Posteriorly it runs out into a well-developed postero-lateral process, from which the concave posterior border runs inwards and backwards to the pointed internal process, which is separated from its fellow in the middle line by a **V**-shaped notch.

The *humerus* (Pl. VII. fig. 2 *b*) of the type specimen is a comparatively short stout bone; it is almost completely ossified, the head being strongly convex, owing to the extension of ossification into the cartilaginous cap. The surface of the head (*h.*) is still much roughened, and bears a number of small prominences, perforated at their summit for the passage of nutritive vessels. This roughened surface is continued on to the upper end of the strong tuberosity (*l.p.*), the form of which is shown in the figure. The outer surface of the process near its upper edge is raised into a series of strong ridges parallel with the long axis of the bone. The ventral surface of the upper end of the shaft is much roughened, and raised into slight prominences for the attachment of muscles. In section the outline of the shaft is a depressed oval, the long axis being in the same direction as the distal expansion. This is not very greatly developed, and is mainly postaxial; the upper and lower surfaces in this region are much roughened, the bone presenting a peculiar fibrous appearance, seen also in other Plesiosaurs, and being perforated by a number of small foramina running obliquely into the substance of the bone towards the shaft.

The distal end bears facets for the radius and ulna: the former of these surfaces is the larger and is nearly flat; it makes a very obtuse angle with the ulnar facet, which is a little concave. Postaxially the cartilage-covered surface turns upwards on to the posterior border of the distal expansion, and there may have been a small postaxial accessory ossicle as in some other forms.

The shape of the *radius* and *ulna* will be best understood from the figure (Pl. VII. fig. 2 b); it will be seen that they unite with one another by a large proximal surface and a smaller distal one, enclosing between them an almost circular opening. The radius (r.) is considerably larger than the ulna (u.), particularly as to the length of its preaxial border. The postaxial border of the ulna bears three facets, which were cartilage-covered in life and may have supported the pisiform and the postaxial accessory ossicle above referred to. Professor Seeley, in describing these specimens, drew attention to the tendency to the elongation of the radius and the transverse extension of the ulna: this character is better marked in another specimen (R. 3698).

Distally the radius and ulna articulate with the radiale, intermedium, and ulnare; the radial and ulnar facets for the intermedium are of about the same size.

In the second specimen (R. 3698) a left *femur* is preserved. In this bone the head is well ossified and convex, its surface being similar to that seen in the humerus described above. The trochanter is large, forming a considerable prominence on the dorsal surface of the upper end of the bone, the ventral surface of which is much roughened for the attachment of muscle; there is also a short but prominent ridge on the postero-superior surface of the shaft. The distal extremity is less expanded than in the humerus, and the expansion is more equally preaxial and postaxial.

The surfaces for articulation with the tibia and fibula are nearly flat, that for the tibia being a little the larger. Postaxially the fibular surface passes up for a short distance on to the posterior border of the expansion. The bone here regarded as the *tibia* is rather longer than broad; it unites posteriorly at its proximal end with the *fibula* by a large facet, but does not seem to have touched it distally. The fibula is broader than long and closely resembles the ulna in form.

R. 1965 (Leeds Coll. 14). A portion of a skeleton including six cervical vertebræ, two dorsals, ten ribs, scapulæ, coracoids, interclavicle, humeri, one radius, ulnæ. Type specimen described and figured by Seeley, *loc. cit. supra*, as *Murænosaurus beloclis*. (Pl. VII. figs. 2–4; text-fig. 70.)

The dimensions (in centimetres) of this specimen are:—

PICROCLEIDUS BELOCLIS.

Vertebræ (Pl. VII. figs. 3-4)	Cervicals.			Anterior dorsals.	
Length of centrum in mid-ventral line	3·2	3·2	3·2	3·2	3·4
Width of anterior face of centrum	4·5	4·3(app.)	4·6	5·3	5·0
Height of ditto	3·0	3·0	..	3·4	3·4
Height to top of neural spine	11·4	11·5	12·5		
Width of middle of neural spine	2·6	2·6	2·6		
Width between ends of transverse processes	..	..	..	10·6	10·7

Shoulder-girdle (Pl. VII. figs. 2, 2 *a* ; text-fig. 70):
   Scapula : greatest length . . . . 18·7
          greatest length from posterior end of median coracoid process to anterior angle . . . . . 10·6
          greatest diameter of coraco-scapular foramen . . 8·8
   Coracoids: greatest length . . . . 27·7
          width of united bones at hinder angles of glenoid cavity . . . . 27·8
          width of combined bones at narrowest . . . . 16·7
          width between outer angles of the postero-external processes . . . . . . (approx.) 14·2

The widths are exaggerated by crushing.

Interclavicle (Pl. VII. fig. 2 *a*):
   greatest length . . . . 7·0
   ,,   width . . . . 3·8
Humerus (Pl. VII. fig. 2 *b*):
   greatest length . . . . 18·3
   diameter of head . . . . 4·9
   greatest width at proximal end . . . . 6·0
   width of shaft at narrowest . . . . 4·4
   ,,   distal expansion . . . . 9·9
Radius (Pl. VII. fig. 2 *b*):
   length in middle line . . . . 5·9
   width at proximal end . . . . (approx.) 5·5
   ,,   distal end . . . . 5·0
Ulna (Pl. VII. fig. 2 *b*):
   greatest length . . . . 4·8
   greatest width . . . . 5·3

R. 3698 (Leeds Coll. 173). Portions of skull and skeleton including basioccipital, quadrate with part of squamosal, a much crushed exoccipital, portions of mandibular rami, atlas, axis, and thirty other cervical vertebræ, many with ribs and neural arches (ribs fused in the anterior part of the neck, free posteriorly), sixteen caudal vertebræ, numerous separate cervical and caudal ribs, ? chevrons, radius and ulna, femora, tibia, and fibula.

The dimensions (in centimetres) of this specimen are :—

Skull :
Length of basioccipital . . . . 2·3
Width of quadrate articulation . . . . 1·4
Mandible (Pl. VII. fig. 1):
Width of articular surface for quadrate . . . . 1·4
Length of postarticular region . . . . (least) 2·1
Length from angle to hinder end of dental series . . . . 7·2

Vertebræ (Pl. VII. figs. 5, 5 a, 5 b, 6)	Atlas and axis.	Cervicals.						
		3rd.	10th.	15th.	20th.	25th.	30th.	35th.
Length of centrum in mid-ventral line	2·7	1·6	2·1	2·6	2·9	3·1	3·1	3·1
Width of posterior face	2·0	2·0	2·5	2·9	3·3	3·7	4·1	4·3
Height of posterior face	1·3	1·3	1·6	2·0	2·4	2·5	2·7	2·7 (app.)
Height to top of neural spine	3·2	..	4·0+	6·0 (app.)	6·2 (app.)	8·5		

	Caudals.					
	Anterior.	Anterior*.	Behind last.		Further back.	
Length in mid-ventral line	2·3	2·2	2·3	2·0	1·8	1·5
Width of posterior face	3·6	3·7	3·2	2·6	2·5	2·0
Height of posterior face	2·4	2·3	2·5	2·1	1·9	1·6
Height to top of neural spine	..	6·5	6·8	4·6	4·6	

Radius : length of outer border . . . . 6·6
" inner border . . . . 3·8
width of proximal end . . . . 5·1
" distal end . . . . 4·9
Ulna : greatest length . . . . 5·2
" width . . . . 5·2
Femur : length . . . . 16·0
diameter of head . . . . (approx.) 3·8
greatest width of proximal end . . . . 5·2
width of shaft at narrowest . . . . 3·5
" distal expansion . . . . 8·6

Picrocleidus sp.

In addition to the skeletons undoubtedly belonging to the type species of this genus, there are remains of a small Plesiosaur which is probably referable to this genus, but differs from the typical form in the shape of the centra of the cervical

---

* First with facets for chevrons.

vertebræ, which are deeper in proportion to their length and width, though otherwise very similar; so far as is known, the shoulder-girdle and humerus are not distinguishable from those of *P. beloclis*. Further material will probably justify the separation of this form as a distinct species, but for the present it need not be named: the specimen (R. 2429) here referred to, including the nearly complete shoulder-girdle and left humerus, with some posterior cervical and the pectoral vertebræ, an anterior dorsal, and some ribs, may be taken as the type.

R. 2429 (Leeds Coll. 41). Portion of a skeleton of an adult individual. The parts preserved include the six posterior cervical vertebræ, two pectorals and the first dorsal, some dorsal ribs; the shoulder-girdle, wanting the interclavicle and the posterior part of the coracoids; the left humerus. The shoulder-girdle, so far as preserved, is closely similar to that of the type specimen of *Picrocleidus beloclis*, and the same may be said of the humerus. The vertebræ, on the other hand, have centra in which the height is greater in proportion to the length and width than in the type species.

The dimensions (in centimetres) of this specimen are:—

Vertebræ	Posterior cervical.	Last cervical.	First pectoral.	Second pectoral.	Anterior dorsal.
Length of centrum in mid-ventral line	3·9	3·7	3·8	3·8	4·0
Width of posterior face of centrum	5·5	6·0	5·9	5·7	5·3
Height of posterior face of centrum	4·3	4·4	4·4	4·4	4·6
Height to top of neural spine.	14·8	?	15·6	15·3	15·4

Shoulder-girdle:
Scapula: greatest length . . . . 22·7
Coracoids: width of the combined bones between the postero-external angles of the glenoid cavity . 29·5
width of united bones at narrowest . . . . . 19·7
antero-posterior diameter of the coraco-scapular foramen . . . . 7·9
transverse diameter of the coraco-scapular foramen 7·9
Humerus: length . . . . 22·5
diameter of head . . . . 5·9
greatest width at proximal end . . . . 8·0
diameter of shaft at narrowest . . . . 5·5
width of distal expansion . . . . 12·6

R. 2739 (Leeds Coll. 33). Portions of the skeleton, including basioccipital with part of basisphenoid and parasphenoid, twenty-six cervical vertebræ mostly with the arches and ribs, two sacrals with the sacral ribs, one anterior caudal, some ventral ribs, both radii, odd paddle-bones, portions of ilia, ischia, and pubes.

In this specimen the fusion of the ribs and neural arches with the centra of the cervical vertebræ is complete, and the same is the case in the single caudal preserved. In the sacrals the ribs remain distinct (text-fig. 71) and it is interesting to note that towards their outer extremities they converge and were in contact or separated only by pads of cartilage; probably three or four sacrals were present. It is not clear what the relationship of the outer ends of the sacral rib was to the ilia, but probably there was no actual contact and the connection was only by ligament.

Text-fig. 71.

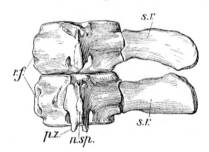

Sacral vertebræ and ribs of *Picrocleidus* sp., from above. ⅔ nat. size (R. 2739). *n.sp.*, neural spine; *p.z.*, posterior zygapophysis; *r.f.*, facets for sacral ribs; *s.r.*, sacral ribs.

The pubes and ischia are very completely ossified, and their median symphysial borders are continuous, the obturator foramina being completely surrounded by bone: the blade of the ischium is rather narrower and more elongated than in *Muraenosaurus*.

The dimensions (in centimetres) of this specimen are:—

Basioccipital: length . . . . (approx.) 2·9
width at lateral processes . . . . 3·7
transverse diameter of occipital condyle . . . . 1·9

Vertebræ	Anterior cervicals.			Middle cervical.	Posterior cervicals.	Sacrals.		Anterior caudal.
Length of centrum in mid-ventral line	2·5	2·9	3·2	4·0	4·0	3·3	3·4	2·9
Width of posterior face of centrum	3·1	3·4	3·7	4·3	5·1	4·8	4·9	4·9
Height of posterior face of centrum	2·3	2·6	2·8	3·5(app.)	4·1	3·4	3·5	3·5
Height to top of neural spine	5·6		7·9	9·5(app.)	13·2 (app.)			

Length of the sacral ribs (text-fig. 71) . . . . 5·7
Ischium: length of expanded portion . . . . 17·5
width of neck . . . . 4·4
„ articular head . . . . 6·8

Genus **TRICLEIDUS**, Andrews.

[Ann. Mag. Nat. Hist. [8] vol. iv. (1909) p. 419.]

Small Plesiosaurs in which the skull is short and broad, with twenty teeth on each side (five on premaxilla, fifteen on maxilla). Pterygoids bear well-developed processes for union with the basisphenoid. Parasphenoid broad and abruptly truncated in front. Quadrate region apparently consisting of two elements (? quadrate and quadrato-jugal). Teeth long, slender, and very sharply pointed; the anterior maxillary teeth enlarged. Neck rather more than three times the length of the skull, and consisting of about twenty-six vertebræ (including the atlas and axis); the centra with strongly concave articular ends, which are much wider than high. The cervical ribs have a prominent anterior angle. In the shoulder-girdle there is a large interclavicle with well-developed, elongated clavicles. The humerus is stout and is not greatly expanded at the distal end, where it articulates with four elements, the radius, ulna, pisiform, and a small accessory postaxial ossicle, probably sometimes wanting. Femur more slender than humerus, and articulating with two bones only.

It has been found necessary to establish this genus for the reception of a small Plesiosaur which differs in some important respects from *Cryptocleidus*, *Murænosaurus*, and other forms with which it has been compared. The chief peculiarities are the possession of well-developed processes of the pterygoids for union with the basis cranii, the presence both of well-developed interclavicle and clavicles, and the distal articulation of the humerus with four elements. Only one species is known at present.

Tricleidus seeleyi, Andrews.

[Plate VIII.; text-figs. 72–77.]

1909. *Tricleidus seeleyi*, Andrews, Geol. Mag. [8] vol. iv. p. 421, text-figs. 1, 2.

Type Specimen.—An imperfect skeleton including the disarticulated bones of the skull (text-figs. 72–75), the mandible (Pl. VIII. figs. 1, 1 *a*), cervical vertebræ (Pl. VIII. figs. 8, 8 *a–d*), pectorals (Pl. VIII. fig. 7), five dorsals (Pl. VIII. figs. 5, 6), and two caudals; numerous dorsal and ventral ribs, coracoids, scapulæ, clavicles, interclavicle (Pl. VIII. fig. 3); fore paddles (imperfect) (text-fig. 77), one pubis; imperfect hind paddles (Pl. VIII. figs. 4, 4 *a*) (R. 3539, Leeds Coll. 39). The pectoral girdle and fore paddle have been figured in the Geol. Mag. *loc. cit. supra.*

The following account of the skeleton in this species is founded on the type and only known specimen.

Skull (text-figs. 72–75).—In the skull here described most of the bones are separate from one another and, in many cases, much crushed and broken; some are missing. Some of the elements are so little distorted that it has been possible to join them together again in their natural position, as in the case of the basis cranii and the pterygoids shown in text-figure 74. The general form of the skull, as a whole, cannot be made out with certainty, but it must have been very similar to that of *Murænosaurus* (text-figs. 46, 47), since the individual bones are for the most part like the corresponding elements in the skull of that genus. In consequence of this general similarity the bones will not be described in detail, except where they present peculiar features.

In the *basioccipital* (text-figs. 73–74) the occipital condyle (*oc.c.*) is more strongly convex, and the pterygoid processes are relatively larger than in *Murænosaurus*. The united *exoccipital* and *opisthotic* (text-fig. 72) are not so stout, and the paroccipital

Text-fig. 72.

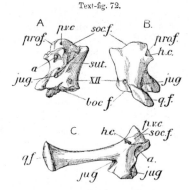

Right exoccipital-opisthotic of *Tricleidus seeleyi*: A, from inner side; B, from outer side; C, from front. (R. 3539, nat. size.)

a., cavity for ampulla of posterior vertical semicircular canal; *boc.f.*, facet for union with the basioccipital; *h.c.*, channel for horizontal semicircular canal; *jug.*, jugular foramen; *pro.f.*, surface for union with prootic; *p.v.c.*, channel for posterior vertical semicircular canal; *q.f.*, facet for quadrate; *soc.f.*, surface for union with supraoccipital; *sut.*, line of juncture between the exoccipital and opisthotic; *XII*, foramen for the XII nerve.

processes, the distal ends (*q.f.*) of which are much expanded, are longer and more slender than in *Murænosaurus*. The line of union (*sut.*) between the exoccipital and opisthotic elements is clearly traceable on the inner face and on the surfaces for union with the basi occipital (*boc.f.*) and supraoccipital (*soc.f.*) bones. The various cavities and foramina for the nerves, blood-vessels, and auditory apparatus are much as in *Murænosaurus* (*cf.* text-fig. 45). The supraoccipital is not known.

The *basisphenoid* (*bs.*, text-figs. 73, 74), as in *Murænosaurus*, consists of a thickened posterior body and an anterior portion, the upper surface of which is deeply hollowed for the reception of the pituitary body (*pit.foss.*). The postero-external angles of the posterior portion of the bone are obliquely truncated by small facets (*pt.f.*¹) for union with the anterior ends of the basisphenoid processes of the pterygoids. The upper anterior angles of the body of the bone bear a pair of facets (*f.*, in text-fig. 73, A) looking upwards and a little outwards, which, judging from comparisons with some recent forms, united with the lower end of the anterior prolongation of the prootics. The posterior surface of the pituitary fossa is perforated on either side by a large foramen for the internal carotids (*i.c.f.*); from the outer anterior border of each of these openings a sharp crest runs outwards and forwards on the cranial surface,

Text-fig. 73.

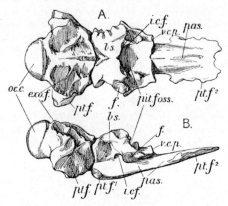

Basioccipital, basisphenoid, and parasphenoid of *Tricleidus seeleyi*: A, from above; B, from right side. (R. 3539, nat. size.)

bs., basisphenoid; *exo.f.*, surface for union with the exoccipital; *f.*, in fig. A facet for (?) lower end of prootic, in fig. B facet for pterygoid; *i.c.f.*, internal carotid foramen; *oc.c.*, occipital condyle; *pas.*, parasphenoid; *pit.foss*, pituitary fossa; *pt.f.*, facet for posterior ramus of pterygoid; *pt.f.*¹, facet for process of pterygoid; *pt.f.*², facet for inner border of palatal ramus of pterygoid; *v.c.p.*, lower cylindrical processes of basisphenoid.

apparently defining the outer side of the pituitary fossa. External to this ridge and in front of the carotid foramen the sides of the bone are produced into short wing-like processes, terminating on facets (*f.*, in text-fig. 73, B), looking outwards and forwards, probably for union with the pterygoids, though this is not certain. The basisphenoid terminates anteriorly in two short vertically-compressed processes (*v.c.p.*) ending in transversely elongated facets for union with the bone or cartilage of the presphenoid region. The ventral face of the basisphenoid is almost completely concealed by the adherent *parasphenoid* (*pas.*), the postero-external angle of which may perhaps take part in the formation of the facets ($pt.f.^1$) for the processes of the pterygoids. The ventral face of the posterior part of the bone is covered with irregular rugosities; further forwards it becomes quite smooth. From its posterior end the bone narrows gradually till just in front of the anterior end of the basisphenoid, when it suddenly widens out, the lateral borders of the widened portion bearing oblique surfaces ($pt.f.^2$) for union with the inner edge of the pterygoids. Anteriorly it terminates abruptly in a thin, sharp, and somewhat concave border. The free portion anterior to the basisphenoid is thin, but is strengthened by the presence of a pair of blunt ridges on the upper surface on either side of the middle line, which is marked by a longitudinal groove. The form and relations of the parasphenoid are shown in text-figs. 73, 74.

The *parietals* are not well preserved; it can be seen that the sagittal crest was short and was high anteriorly. Posteriorly the bones send off short stout lateral processes which unite with, and are to some extent overlapped by, the squamosals, while anteriorly they widen out and are separated in the middle line by the large parietal foramen, of which they form nearly the whole border. Their anterior edges unite internally with the posterior ends of the frontals, which close the front of the pineal foramen, and externally to these there is a broad sutural surface which, judging from the structure in *Murænosaurus*, united with the postfrontals. On the ventral surface of the united parietals there is posteriorly a broad facet, looking backwards and downwards, for union with the supraoccipital. In front of this the cranial surface of the bones is at first nearly flat, becoming more and more concave forwards as it passes into the posterior wall of the pineal opening, which is bordered laterally by strong ridges. On the posterior border of the anterior expansion of the parietals there is a pit-like facet, probably for the reception of the upper end of the *columella cranii*.

The *frontals*, so far as they are preserved, are much like those of *Murænosaurus*. The nasals are unknown, and indeed the whole of the bones of the upper surface of the skull are badly preserved or wanting altogether; it appears, however, that the external nasal opening was large.

The *maxillæ* each bear fifteen teeth, of which the third and fourth are the largest. behind them there is a gradual decrease in size to the hinder end of the series. The facial (suborbital) plate of the bone is large, thin, and concave superiorly. The palatal portions of the bone are small. The sutural surface for union with the premaxillæ is straight and nearly at right angles to the long axis of the skull.

The *premaxillæ* are almost triangular in outline and are prolonged back in the middorsal line into short facial processes. The facial surface is covered with pits and strong ridges. Each bears five teeth—the first small, the second, third, and fourth large, the fifth again small.

The *vomers* are closely united into a triangular plate of bone, and their upper surface bears a deep median groove, at the bottom of which traces of the original separation of the two bones can be seen. At the posterior end of the groove are two deep depressions which seem to have received the anterior ends of the pterygoids.

Text-fig. 74.

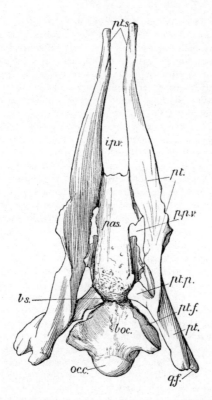

Basis cranii and pterygoids of *Tricleidus seeleyi*, from below. (R. 3539, nat. size.)

boc., basioccipital; *bs.*, basisphenoid; *i.p.v.*, interpterygoid vacuity; *oc.c.*, occipital condyle; *pas.*, parasphenoid; *p.p.v.*, posterior palatine vacuity; *pt.*, pterygoid; *pt.f.*, facet for posterior ramus of pterygoid; *pt.p.*, basisphenoid process of pterygoid; *pt.s.*, sutural surfaces of anterior ends of pterygoids, the bones being separated through distortion; *q.f.*, facet for union with quadrate.

Laterally the bones are thickened and bear sutural surfaces looking upwards and outwards; these are mainly for union with the palatine plates of the premaxillæ, but posteriorly probably also with the palatines. The posterior portion of the lateral border has a smooth slightly concave edge, presumably the inner side of the internal narial opening. The palatal face of the combined bones is gently convex from side to side.

The *pterygoids* (text-fig. 74) are peculiar in several respects, differing considerably from those of *Murænosaurus*. The outer (lateral) ramus seems to have been very short or absent, the transverse bone having united directly with the lateral border of the body of the bone, where a facet for its reception can be seen. The main body of the bone consists of a comparatively thin palatal portion and a thickened and deepened posterior bar, which at its posterior end unites with the quadrate. The posterior portion of the palatine region is narrow and is strengthened by a dorsal ridge-like thickening, which is, in fact, the anterior prolongation of the thick quadrate region; the dorsal surface of this ridge, which dies away opposite the surface for the transverse bone, bears a roughened facet, probably for union with the lower end of the *columella cranii*. In front of this the palatal plate widens out and its ventral (oral) surface is marked by a series of slight longitudinal grooves. Anteriorly the bone narrows and becomes vertically compressed; in this region it was probably in contact in the middle line with its fellow of the opposite side for some distance (*pt.s.*); behind this point backwards as far as the anterior edge of the parasphenoid the pterygoids are separated by a median interpterygoid vacuity (*i.p.v.*). Anteriorly the extremities of the two bones fitted into the pits on the posterior border of the vomers referred to above. The most peculiar feature of the pterygoids in this genus is the presence on each of them of a long process (*pt.p.*) for union with the postero-external angles of the basisphenoid (and perhaps, to some extent, of the parasphenoid). These processes, which are directed forwards and inwards, arise at the junction of the palatal and quadrate regions of the pterygoids; they are compressed from before backwards and their posterior face is nearly flat, the anterior being convex from above downwards; they terminate in flat oblique facets which fit against the corresponding facets on the basis cranii as above mentioned. The quadrate region of the pterygoid is short, stout, and somewhat compressed from side to side; its outer face is convex, while its inner bears a large flat facet for union with the pterygoid tuberosity of the basioccipital; its posterior end is bifurcated and has an irregular sutural surface for the inner edge of the quadrate, with which it seems to have united very firmly.

The *palatines* and *transverse bones* are not completely known.

The *quadrate region* (text-fig. 75) presents some remarkable features. The articular surface, which in other Plesiosaurians examined is formed entirely by the quadrate, is in this case apparently constituted by two distinct elements. It is just possible that this division may be the result of fracture; but if so, this has occurred symmetrically on both sides and the resulting surfaces look like the faces of a suture. Of the two parts the inner (*q.'*) is the smaller and is narrow from side to side; its inner border bears a deeply hollowed and roughened surface for union with the posterior end of the pterygoid; distally it bears the surface for the inner third of the articulation of the mandible (*cond.*), this surface being strongly convex from before backwards and slightly so from side to side. The outer border, where union with the larger outer

Text-fig. 75.

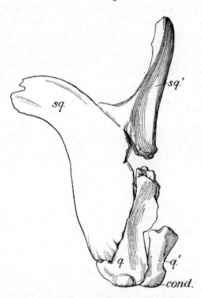

Left squamosal and quadrate of *Tricleidus seeleyi*, from outer side   (R. 3539, nat. size.)

cond., condyle for mandible; *q.*, outer portion of quadrate (? quadrato-jugal); *q.'*, inner portion of quadrate; *sq.*, zygomatic bar of squamosal; *sq.'*, parietal bar of squamosal.

element (*q.*) takes place, is irregularly roughened, the two uniting in a partly overlapping suture. The outer two-thirds of the articular surface for the mandible is borne by the second, larger element; this surface is strongly convex from before backwards, very narrow at the junction of the two elements, but widening towards its outer side. Externally there is a pointed process, and on the upper edge of this, and on the long straight border above it, is a clearly defined sutural surface for union with the lower ramus of the squamosal (*sq.*).

There seem to be two possible explanations of this peculiar structure of the quadrate region: one, as already mentioned, that the two quadrates have been fractured in an exactly similar way, the other that there are really two distinct elements. In the latter case the smaller inner bone (*q.'*) uniting with the pterygoid would be the true quadrate, while the larger outer one overlapped by the squamosal would be a quadrato-jugal. This latter would bear the greater part of the articular surface for the mandible, a condition which, so far as I am aware, is never found in other reptiles, though in *Sphenodon* the quadrato-jugal does seem to enter into the formation of the outer border of the articulation. The absence of the division in any specimens, some quite young, of other species, favours the idea that a symmetrical fracture has occurred, but, nevertheless, it seems well to suggest the other explanation also.

The *squamosal* (text-fig. 75) is a large triradiate bone as in *Murænosaurus*, but there are some differences of detail. The dorsal bar (*sq.'*) running up to meet the squamosal process of the parietal is comparatively slender; at its upper end it is slightly enlarged and bears a deeply pitted surface for union with the parietal. In this genus the dorsal rami of the two squamosals do not seem to have met in the middle line over the parietals. Ventrally the bar widens out and the ridge forming its upper border bifurcates, the anterior arm becoming continuous with the upper edge of the anterior ramus. This is thin and broad, its upper margin being convex, the ventral concave, and passing posteriorly into the anterior edge of the broad ventral (quadrate) ramus. The posterior border of this latter has on its inner face an extensive sutural surface for union with the quadrate (or quadrato-jugal), which it overlaps to a considerable extent. The anterior end of the zygomatic process (*sq.*) probably united both with the jugal below and the lower edge of the postorbital above.

The remainder of the skull is represented by mere fragments, which it has not been found possible to piece together.

The *mandible* (Pl. VIII. figs. 1, 1 *a*) is very well preserved in the type specimen. In its general structure it is closely similar to that of *Murænosaurus*, the same fusion between the constituent bones having occurred. It is rather stouter in proportion to its length than in *Murænosaurus*, at least in the tooth-bearing region; the coronoid angle, which seems to be formed by the hinder end of the splenial, is very little developed; the symphysis is short.

The *teeth* (Pl. VIII. figs. 1, 2) in both the upper and lower jaws are long, slender, curved, and terminating in very sharp points. The enamel at the lower part of the crown is raised into fine nearly parallel ridges; these mostly die away towards the point, only one or two continuing to the actual tip.

Vertebral Column (Pl. VIII. figs. 5–8).—In the length of the neck and in the structure of the individual vertebræ this genus approaches *Cryptocleidus* and differs widely from

Murænosaurus. The number of cervicals (assuming that the series collected is complete, as it appears to be) is only twenty-five, or even less than in *Cryptocleidus*, and only little more than half the number in *Murænosaurus* (44). The atlas and axis (Pl. VIII. fig. 8) are similar to those of *Murænosaurus* figured above (text-fig. 49), except in some details: thus the hypapophysial ridge, though not so strongly developed in front, extends quite to the posterior end of the axis. The rib of the atlas (Pl. VIII. fig. 8, $r.^1$) is much longer than in *Murænosaurus*, though still much smaller than that of the axis ($r.^2$); the neural arches (*at.a.*, *ax.a.*) are lower. The other members of the cervical series are characterised by the possession of short centra (if anything, a little shorter than in *Cryptocleidus*), the articular surfaces of which are considerably broader than high and rather strongly concave (Pl. VIII. figs. 8, 8 *a*, 8 *b*, 8 *c*, 8 *d*) with a deep central pit; the concavity is bordered by a well-developed rounded border. The upper surface is slightly concave beneath the neural canal. The lateral surfaces of the centra above the facets for the cervical ribs are concave both from before backwards and above downwards; the ventral face is, as usual, perforated by a pair of nutritive foramina separated by a narrow ridge.

The neural spines are relatively narrower from before back than in *Cryptocleidus*, and towards the hinder part of the neck also higher. The cervical ribs have a well-developed anterior angle, a remnant of the hammer-head shape of the ribs in some of the earlier Plesiosaurs: this angle is less developed or absent altogether in *Cryptocleidus*, but is well-marked in *Picrocleidus* (see Pl. VII. fig. 5, *a.p.*). Behind the cervical series there are two vertebræ in which the rib is borne partly on the centrum and partly on the incipient transverse process of the arch; these may be called the pectorals. Behind these, in the type specimen, follow five dorsals, all that remain of the rest of the vertebral column except two small posterior caudals. In the pectorals and anterior dorsals (Pl. VIII. figs. 5, 6, 7) the centrum becomes more nearly circular in outline as we pass backwards, and the nutritive foramina ascend to its sides, where they open at the bottom of a well-defined depression. At the same time the ventral surface becomes flatter, so that at the fifth (the last preserved) dorsal it is only slightly convex from side to side (Pl. VIII. fig. 5); the articular faces continue to be deeply concave. The transverse processes increase rapidly in length as they pass backwards; the neural spines are high and relatively narrow.

Of the two small posterior caudals the anterior one is considerably larger at its anterior than at its posterior end; the neural arch seems to have been very small and low, and was confined to the front of the centrum. On the sides there are prominent facets for the caudal ribs. On the ventral surface there is on either side a strong rounded ridge cut away obliquely both in front and behind by the chevron facets; both between and above these ridges the body of the centrum is concave from side to side. The second still smaller caudal also bears facets for ribs and chevrons.

The *shoulder-girdle* (Pl. VIII. fig. 3; text-fig. 76) is typically Elasmosaurian, the scapulæ and coracoids meeting in a long continuous median symphysis. The structure of the clavicular arch in this genus distinguishes it from its contemporaries, in that both the clavicles (*cl.*) and the interclavicle (*i.cl.*) are well developed. The *interclavicle* is a transversely elongated oval bone, the outer convex borders of which are produced into a number of irregular projections; the anterior border is slightly concave in the middle line, the posterior nearly straight. The peripheral portions of the bone are thin, but in the middle line, especially at about the junction of the

Text-fig. 76.

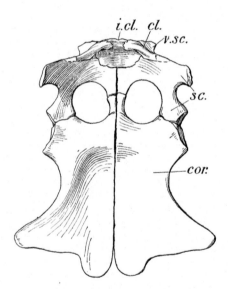

Shoulder-girdle of *Tricleidus seeleyi*, from above. (Type specimen, R. 3539, ⅙ nat. size.)
cl., clavicles; *cor.*, coracoid; *i.cl.*, interclavicle; *sc.*, scapula; *v.sc.*, ventral ramus of scapula.

middle and anterior thirds of the bone, there is a considerable thickening, which appears on the inner (visceral) surface as a slight longitudinal elevation, on either side of the posterior portion of which the surface is gently concave. These concavities are bounded anteriorly by a transverse ridge, against which the inner ends of the clavicles (*cl.*) fit. The ventral face of the interclavicle is gently convex and its surface is slightly pitted and roughened, especially towards the periphery, where the bone seems to have a fibrous structure. The *clavicles* (*cl.*) are elongated bones, consisting of a thickened central axis which at their median ends widens out and bears, both on its anterior and posterior margins, thin expansions which fit against the visceral face of the anterior part of the interclavicle. The outer ends of the clavicles are not well preserved, but probably they terminated in points resting against the anterior borders of the scapulæ as in *Cryptocleidus*; the inner ends of the clavicles do not seem to have met in the middle line. From the above account it will be seen that in this genus the clavicular arch is in many respects intermediate in structure between that of *Cryptocleidus* and of *Murænosaurus*.

The general form of the *scapula* will be best understood from the figure (text-fig. 76, *sc.*). The ventral bar is large, though not so much expanded as in some species of *Murænosaurus*, e. g. *M. platyclis*. The symphysis is thickened posteriorly, but thins out towards the front, terminating in a point beyond which the median borders of the thin anterior portions diverge from one another, leaving a V-shaped notch. This is concealed above by the overlying interclavicle, which rests in the depression formed by this thin anterior portion of the scapulæ. The anterior edges of the dorsal rami are continued on to the visceral face of the ventral rami as ridges, against the front of which the clavicles seem to have fitted; and a roughened surface on the anterior edge of the dorsal ramus near its lower end probably marks the point of attachment of the outer end of the clavicle. Above this surface the anterior edge of the dorsal ramus becomes thin and sharp; the whole is directed strongly backwards. The posterior (articular) arm of the scapula is closely similar to the same part of the Murænosaur scapula.

The general form of the *coracoid* is shown in text-fig. 76. The bone, as a whole, is very thin, especially in its posterior median portion. Anteriorly, on a line connecting the two glenoid cavities of the conjoined bones, there is a great thickening, especially towards the middle line, where the inner surface of the thickened portion bears the roughened symphysial facet for the union of the two sides; this symphysis is continued forwards between the median prolongations of the coracoids and becomes continuous in front with the scapular symphysis; the thin posterior portions of the coracoids also unite in the middle line for a considerable distance. The visceral surface of this region of the coracoids is deeply concave. The prominent postero-lateral processes terminate in a slightly concave surface, showing that in life they were tipped with cartilage. The surface for articulation with the scapula is triangular in form and is covered with rugosities: the glenoid surface is three-quarters of an oval; it is nearly smooth, showing that ossification was complete, as it is found to be in other parts of this skeleton. The anterior prolongations of the coracoids for median union with the scapulæ are not quite symmetrical, that on the right side being rather longer than the other; in the shoulder-girdle of the type specimen of *Murænosaurus platyclis* the same peculiarity is noticeable.

The *fore limb* (text-fig. 77) is peculiar in several respects and differs considerably from that of *Cryptocleidus* and of *Muraenosaurus*; its most striking characteristic is that the humerus articulates distally with four bones—three, the radius (*r.*), ulna (*u.*), and pisiform (*p.*), being large, the fourth a small postaxial accessory ossicle (*a.*).

The *humerus* (*h.*) is short and stout; the head is rounded in outline and convex; at its upper anterior border it is continuous with the surface of the strongly developed tuberosity (*l.p.*). This is bounded both in front and behind by strong ridges, which extend down a little on the shaft; its upper surface is flattened, and a little

Text-fig. 77.

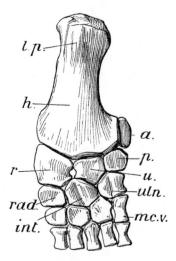

Left fore paddle (imperfect) of *Tricleidus seeleyi*, from above. (Type specimen, R. 3539, about ⅓ nat. size.)
a., postaxial accessory ossicle; *h.*, humerus; *int.*, intermedium; *l.p.*, tuberosity;
mc.v., fifth metacarpal; *p.*, pisiform; *r.*, radius; *rad.*, radiale; *u.*, ulna; *uln.*, ulnare.

below its upper border there is a well-marked rugosity for the attachment of muscle. The stout shaft is oval in section; its anterior border bears a roughened ridge, and the upper part of its ventral and ventral-posterior surface is roughened for muscle-attachments. Distally the bone is expanded and compressed from above downwards. The facet for the radius is the largest, that for the pisiform the smallest; the surface for the accessory ossicle is situated entirely on the postaxial border nearly parallel with the long axis of the bone.

The *radius* (*r.*) is considerably larger than the ulna; it articulates with the humerus by a long oblique and nearly straight facet; its outer border is thin and rounded, while the inner bears two facets for union with the ulna, one proximal, the other distal, the two bones being separated in the middle by a rounded foramen. Distally the radius articulates with the radiale by a long facet, and behind this, and making an obtuse angle with it, there is a short facet for the intermedium. The *ulna* (*u.*) has a nearly flat facet for the humerus; anteriorly it unites with the radius at its proximal and distal ends, being separated from it in the middle by the foramen mentioned above. Distally it joins the intermedium and ulnare, the facets for which make with one another an angle a little larger than a right angle; postaxially it unites with the *pisiform* (*p.*). This bone articulates proximally with the humerus, anteriorly with the ulna, and distally with the ulnare; it also has a short facet on its posterior border for contact with the *accessory ossicle* (*a.*). This is an elongated bone which unites by a long facet with the postaxial border of the distal expansion of the humerus, and, as above noted, its lower end is in contact with the pisiform.

The form and arrangement of the *carpal bones* will be best understood from the figure (text-fig. 77). There are three carpals in each row, those in the distal rows being small; the fifth metacarpal (*mc. V*), as usual in these reptiles, articulates with the ulnare and has on its preaxial face at the proximal end a facet for contact with the third distal carpal.

The other metacarpals articulate proximally as follows:—metacarpal I. with the first distal carpal only; metacarpal II. with the second carpal; metacarpal III. partly with the second and partly with the third carpal; metacarpal IV. with the third carpal only.

Of the pelvis only a *pubis* is known. This is a broad plate of bone, the length and breadth of which are almost equal, in this respect resembling the pubis of *Murænosaurus* rather than that of *Cryptocleidus*, which is considerably wider than long. The greater part of the bone is thin, but it thickens towards the symphysial border, the symphysial surface being rather deep, especially towards the front. The articular region is also much thickened and bears two subequal facets, one for union with the ischium, the other forming the anterior part of the acetabulum; as usual, there is no contact with the ilium. The anterior edge of the bone is somewhat irregular in outline, not being evenly convex as in some Plesiosaurs; the visceral surface is slightly concave, the outer surface flat or slightly convex.

The *femur* (Pl. VIII. figs. 4, 4 *a*) is a much more slender bone than the humerus, but of about the same length. The head (*h.*) is convex, and the trochanter (*tr.*, fig. 4 *a*) is very strongly developed, its upper surface being continuous with that of the head. Anteriorly it is bounded by a strong ridge, but the ridge forming its posterior border is not so strongly marked. The ventral face of the upper half of the shaft bears a much roughened surface (*m.r.*) for muscle-attachment, and most of the posterior face of the shaft is likewise rugose. Distally the bone is expanded to a smaller degree than the humerus, and so far as is known it articulated with two bones only, the tibia and fibula.

The *tibia* (Pl. VIII. fig. 4, *t.*) is somewhat larger than the fibula; its surface for union with the femur is gently convex, as also is its preaxial border. Postaxially it unites with the fibula at its upper and lower ends, the two bones being separated in the middle by a rounded opening; distally it has a long gently concave surface for the tibiale, and making an obtuse angle with this a short facet for the intermedium.

The *fibula* (Pl. VIII. fig. 4, *f.*) has a long straight proximal facet for the humerus; anteriorly it unites proximally and distally with the tibia, distally it unites with the intermedium in front, and behind has an oblique facet, which no doubt carried the fibulare. The rest of the hind paddle is not known.

R. 3539 (Leeds Coll. 39). Disarticulated bones of skull (text-figs. 72–75), mandible (Pl. VIII. fig. 1), cervical (Pl. VIII. fig. 8), pectoral (Pl. VIII. fig. 7), five dorsal (Pl. VIII. figs. 5, 6) and two caudal vertebræ; numerous dorsal and ventral ribs; shoulder-girdle (text-fig. 76), including coracoids, scapulæ, clavicles and interclavicle (Pl. VIII. fig. 3); fore paddles (text-fig. 77) (incomplete); one pubis; hind paddles (Pl. VIII. fig. 4) (incomplete). Type specimen described and figured in Ann. Mag. Nat. Hist. [8] vol. iv. (1909) p. 419.

The dimensions (in centimetres) of this skeleton are:—

Skull (text-figs. 72–75):
```
Length of basioccipital . . . . . . . . . . . . . .   3·4
Width of basioccipital at lateral processes . . . . . .   4·2
    "     occipital condyle . . . . . . . . . . .   1·9
Length from condyle to anterior end of parasphenoid . . .   8·3
Approximate length of pterygoids . . . . . . . . . .  15·5
Extreme length of mandible . . . . . . . . . . . .  25·0
Width of articular surface for quadrate. . . . . . . .   2·4
Length of symphysis . . . . . . . . . (approx.)   3·5
Approximate length of crown of largest mandibular tooth .   2·9
Width at base of largest mandibular tooth . . . . . .   0·6
```

Vertebræ (Pl. VIII. figs. 5–8)	Atlas and axis.	Second cervical.	Sixth cervical.	Twelfth cervical.	Eighteenth cervical.	Twenty-fourth cervical.	First pectoral.	Dorsal.
Length in mid-ventral line .	4·0	2·3	2·3	2·8	3·0	3·1	3·4	3·7
Width of anterior face of centrum	2·0	2·5	2·8	3·5	4·1	4·4	5·0	4·9
Height of anterior face of centrum	2·1	2·2	2·3	2·8	3·4	3·5	3·8	4·3
Height to top of neural spine	..	5·2	5·7	7·4	9·7	10·5	13·0	13·7

Shoulder-girdle (Pl. VIII. fig. 3; text-fig. 76): total length in
  middle line . . . . 53·0
Interclavicle: length in middle line . . . . 6·4
  „   greatest width . . . . 13·8
  „   „   thickness . . . . 1·4
Clavicle: anterior width . . . . 2·7
Scapula: length of symphysial surface . . . . 10·3
  „ from anterior angle to angle at union of glenoid
       and coracoid surfaces . . . . (approx.) 24·0
  width from symphysis to tip of dorsal ramus (app.) 24·0
  „ of glenoid surface . . . . 3·9
Coracoid: greatest length . . . . 39·0
  „   width at hinder angle of glenoid cavity . . 18·0
  width at narrowest point . . . . 13·1
  „   posterior expansion . . . . 23·5
  greatest depth of symphysial surface . . . . . . 6·1
Antero-posterior diameter of coraco-scapular foramen . . . 10·2
Fore limb (text-fig. 77):
Humerus: length . . . . 20·7
  diameter of head . . . . 6·3
  greatest width of upper end . . . . 7·9
  width of shaft at narrowest . . . . 5·3
  „   distal expansion . . . . 11·8
Radius: length of preaxial border . . . . 5·1
  „   proximal end . . . . 5·6
  „   distal end . . . . 5·0
Ulna: greatest length . . . . 3·5
  „   width . . . . 4·1
Pisiform: width . . . . 3·6
  length . . . . 3·4
Pubis: greatest length . . . . 25·0
  „   width . . . . 23·0
  depth of symphysis . . . . 3·3
  length of surface for ischium . . . . 4·3
  „   acetabular surface . . . . 5·0
  least width of neck . . . . 7·7
Hind limb:
Femur (Pl. VIII. fig. 4):
  length . . . . 21·6
  greatest width at proximal end . . . . 7·8
  width of head . . . . 6·9
  „   shaft at narrowest . . . . 4·3
  „   distal expansion . . . . 11·3
Tibia (Pl. VIII. fig. 4): length . . . . 3·6
  width . . . . 5·6
Fibula (Pl. VIII. fig. 4): length . . . . 3·5
  width . . . . 5·4

Genus CRYPTOCLEIDUS, Seeley.

[Proc. Roy. Soc. vol. li. (1892) p. 145, as a subgenus of *Murænosaurus*.]

Plesiosaurs in which the skull is relatively small, about a quarter the length of the neck in the adult. Neck consisting of about 32 cervical vertebræ, the centra of which are short, with oval articular faces which are deeply concave, at least in old individuals. There are two or three pectoral vertebræ and 21 or 22 dorsals. There seem to have been three or four sacrals, the ribs of which are thickened and converge towards their outer ends; the number of caudals is uncertain, but probably was about thirty; the posterior caudals diminish in size rapidly. Cervical ribs with single heads, not much flattened, and in some cases with a fairly prominent anterior angle. The ventral ribs forming a strong plastron, each transverse row consisting of a median and three pairs of lateral elements. Shoulder-girdle of the true Elasmosaurian type, the scapulæ having an extensive median symphysis continuous with that of the coracoids. Coracoids with prominent postero-lateral processes in the adult. The clavicular arch consisting of two triangular clavicles meeting in median symphysis: in some cases with a rudimentary interclavicle interposed between them posteriorly (text-fig. 88); the clavicular arch, except for its extreme anterior edge, lying entirely on the visceral surface of the ventral bar of the scapulæ in the adult. The humerus greatly expanded distally; radius large, with an elongated anterior border, so that the axis of the expanded portion of the paddle makes a slight angle with that of the humerus. In the pelvis the pubis short in proportion to its width; the femur not greatly expanded distally.

Only a single species of this genus is recognised from the Oxford Clay of Peterborough.

Cryptocleidus oxoniensis, Phillips, sp.

[Plates IX. & X.; text-figs. 78-94.]

1871. *Plesiosaurus oxoniensis*, Phillips, Geology of Oxford, etc. p. 307.
1871. *Plesiosaurus eurymerus*, Phillips, op. cit. p. 315.
1888. *Plesiosaurus oxoniensis*, Lydekker, Geol. Mag. [3] vol. v. p. 352.
1889. *Cimoliosaurus eurymerus*, Lydekker, Catal. Foss. Rept. Brit. Mus. pt. ii. p. 205.
1889. *Cimoliosaurus oxoniensis*, Lydekker, tom. cit. p. 209.
1892. *Plesiosaurus durobrivensis*, Seeley, Proc. Roy. Soc. vol. li. pp. 132-134 (for young).
1892. *Cimoliosaurus eumerus*, Seeley, Proc. Roy. Soc. vol. li. p. 145.
1892. *Cryptocleidus platymerus*, Seeley, tom. cit. p. 145.
1895. *Cryptocleidus oxoniensis*, Andrews, Ann. Mag. Nat. Hist. [6] vol. xv. p. 335.

Type Specimen.—Cervical, dorsal, and caudal vertebræ described and figured by Phillips in the 'Geology of Oxford, etc.,' pp. 307-309, figs. cxiii.-cxv. Phillips also ascribes to this species a shoulder-girdle (figured and described as a pelvis—*op. cit.* p. 310, fig. cxvi.) and a paddle (*op. cit.* p. 312, fig. cxvii.), both of which probably belong to a species of *Murænosaurus*.

Phillips, in his original account of the cervical vertebræ which must be regarded as the type specimens, simply states that they are biconcave with narrow tumid interforaminal space, while his figures and measurements show that the centra are short; the vertebræ figured, however, are not anterior examples as Phillips supposed, but from some little distance back in the neck. The dorsals and caudals described are not definitely stated to have been associated with the cervicals, and, as above noted, the shoulder-girdle and paddle are those of a Murænosaur, consequently the determination of the species must rest entirely on the cervical vertebræ. Comparison of these with the corresponding vertebræ of the commonest type of Plesiosaur from the Oxford Clay of Peterborough, shows such similarity of form that both are clearly of the same species, and therefore the name *Cryptocleidus oxoniensis* is applied to them. Lydekker employed the name *Cimoliosaurus oxoniensis* for the smaller individuals of this type, while he called the larger *C. eurymerus*, a name which had been given by Phillips to a large broad paddle from the Oxford Clay of Bedford associated with vertebræ similar to those of *C. oxoniensis*. The series of specimens in the Leeds Collection tends to show that, as Lydekker himself suggested, these two forms are probably only a single species, and that the difference in size and form are merely the result of increased age. Since, however, the ossification of the shoulder-girdle is completed while the individual is considerably smaller in some cases than in others, it is possible that the difference may be a sexual one, as I have already suggested in a paper on the development of the shoulder-girdle (Ann. Mag. Nat. Hist. [6] vol. xv. (1895) p. 333). The differences in the form of the limb-bones which led Professor Seeley to establish his species *Cryptocleidus platymerus* for a specimen (Leeds Coll., R. 2412) in this collection, are probably due entirely to the advanced state of ossification that had been reached in this case.

One of the most important points about this species is, that remains of individuals of all ages are common in the Peterborough deposits, and are easily distinguished from the other forms. From these specimens it has been possible to make out the history of the development of several parts of the skeleton, notably that of the shoulder-girdle (see Ann. Mag. Nat. Hist. [6] vol. xv. (1895) p. 333).

The description of the skeleton given below is founded mainly on the almost perfect adult skeleton (R. 2860) which has been mounted in the Gallery of Fossil Reptiles; reference will also be made to other specimens, especially to the mounted skeleton of a young individual (R. 2417), a brief account of which has been published in the Geological Magazine (1895), p. 241.

Skull (Pl. IX.).—The skull, so far as known, is very similar in its structure to that of *Murænosaurus*, and will be described mainly by pointing out such differences as occur. Unfortunately the material available for description is very imperfect, all the specimens being much crushed and wanting many important parts.

The *basioccipital* (*b.oc.*, Pl. IX. figs. 1, 2, 4, 5) is closely similar to that of *Murænosaurus*, as will be seen from the figures. The occipital condyle is perhaps a little broader in proportion to its height, and its articular surface is continued quite up to the facet for the exoccipital, there being no trace of a neck to the condyle such as is seen in most specimens of the basioccipital of *Murænosaurus*; the basipterygoid (lateral) processes (*pt.p.*) are a little more rounded in section than in that genus.

The *basisphenoid* (*b.sp.*, Pl. IX. figs. 2, 5) does not differ in any important respects from that of *Murænosaurus* or *Tricleidus*. The facet marked *f.* in text-fig. 73, A, is here very well developed, and probably received the anterior lower angle of the pro-otic. The ventral face is extensively overlapped by the parasphenoid in the usual way. In the young skull (Pl. IX. fig. 5) the basisphenoid shows evidence of its original ossification from two centres, since it is deeply notched posteriorly in the middle line and there is also a large vacuity beneath the pituitary fossa. This opening corresponds to the original space between the trabeculæ, through which the pituitary body was connected with the pharynx; possibly, even in the young, it was partly closed by the posterior end of the parasphenoid, and it corresponds in position to the pit occurring in some specimens of the basisphenoid of *Murænosaurus* (see Pl. III. fig. 1).

The structure of the *exoccipital-opisthotic* (*ex.op.*, Pl. IX. figs. 1, 1 *a*, 4, 4 *a*) differs very little from that seen in *Murænosaurus*; the paroccipital process is rather shorter, and in the oldest specimen available for comparison (R. 2860) the fossæ for the ampulla of the posterior vertical semicircular canal and for the canals themselves are larger and more open (see Pl. IX. fig. 1). In the young specimen (R. 2417) the exoccipital and opisthotic are united above the jugular foramen (*jug.*), their line of union being still clearly visible; but below that opening they are still separate (Pl. IX. fig. 4), each terminating in a distinct facet for union with the basioccipital. In this specimen also the posterior face of each exoccipital bears a small facet (*f.*) somewhat resembling a zygapophysis; this probably indicates that a well-developed pro-atlas was present, since the form of the neural arch of the atlas is such that there certainly was no point of contact between it and the exoccipitals. The *supraoccipital* (*soc.*) is a high arch curving forwards, the occipital surface being concave from above downwards and continuous with the posterior face of the postero-lateral processes of the parietals (Pl. IX. fig. 1 *a*). In front the supraoccipital bears a facet looking downwards and forwards for union with the pro-otic, and the inner side of this surface is deeply channelled for the reception of the upper portion of the posterior vertical semicircular canal (*p.v.c.*). Between the pro-otic facet and the junction with the parietal the border of the supraoccipital is sharp and concave, and clearly did not unite with any other element.

The *parietals* (*par.*) widen out posteriorly into lateral processes, on which are the sutural surfaces overlapped by the upper ends of the dorsal processes of the squamosals (*sq.*[1]), which in this genus do not appear to have met in the middle line, so that the actual vertex of the skull is formed by the parietals alone. In front of the lateral process the united parietals rise into a high sagittal crest, widening out again at the pineal foramen, of which they form, at least the greater part of and possibly all, the margin. From an examination of this region in the young skull it seems probable that the parietals completely surround the pineal foramen at its inner (cranial) end, while on the outer surface of the skull a portion of its anterior margin may be formed by the overlapping frontals. In front of the pineal foramen the *frontals* widen out considerably and unite at their outer ends with the postfrontals; in front of this they form the roof of the orbit, but their relations with the bones further forwards is unknown, all the specimens being very imperfect in the rostral region.

The *squamosal* is of the usual triradiate form (Pl. IX. fig. 3); the slender dorsal rami (*sq.*[1]) run up to the lateral processes of the parietals, with which they unite, overlapping them on the upper and probably also on the lower surface, but not meeting one another in the middle line. Judging from the appearance of this portion of the squamosal in the young specimen, it seems possible that it may have originally ossified from a distinct centre and may therefore represent a supratemporal. The ventral ramus unites closely with the quadrate, down the outer side of which it sends a long process. The anterior (zygomatic) ramus (*sq.*) is thin and broad; its upper border is strongly convex, the lower concave; anteriorly it bears a sutural surface for union with the jugal and presumably also with the postorbital.

The *quadrate* (*q.*) is a large bone; its anterior face is concave from side to side, its posterior face convex or flat. The upper end is embraced by the squamosal, while at its lower end it bears the broad articular surface for the mandible. This surface is convex from before backwards, and is imperfectly divided into a larger outer and a smaller inner convexity. There is no trace of any division of this bone into two elements. The *premaxillæ* seem to have borne six teeth each; the muzzle was probably rather more pointed than in *Murænosaurus*.

The *mandible* (Pl. IX. figs. 6, 7) is somewhat more slenderly built than in *Murænosaurus*, and the symphysis (*sym.*) seems to have been somewhat shorter, otherwise the structure is similar. There is no trace of any division between the articular and surangular, even in the youngest specimens. The postarticular region is relatively smaller and especially shorter than in *Murænosaurus*. There are 25 or 26 teeth on either side.

The *teeth* (Pl. IX. fig. 7) are long, slender, and very sharply pointed; the enamel is smooth, except for a few fine ridges confined to the lower part of the inner side of the crown. In *Murænosaurus* (Pl. III. figs. 4–6), on the other hand, the whole of the crown except the tip is covered with fine longitudinal ridges; on the anterior and the posterior side there is a main ridge, more prominent than the rest and continuous almost to the top of the crown, forming in some cases a very slightly marked keel. The rest of the surface (Pl. III. fig. 4 *a*) is covered with fine ridges of varying length, running in a generally longitudinal direction and often anastomosing. The crown is circular or nearly circular in section throughout its length; in *Cryptocleidus* the crown is sometimes slightly compressed, so that in section it is oval.

The roots of the teeth in both genera are very long and circular in section; their surface is smooth and the pulp-cavity is large.

Vertebral Column.—The *atlas* and *axis* (text-fig. 78) are very similar in structure to those of *Murænosaurus*, but differ in some details. As in the case of the other cervicals, the centra of these vertebræ, particularly that of the axis, are shorter than in *Murænosaurus*. In the formation of the cup for the occipital condyle the bases of the lateral pieces of the neural arch of the atlas (*at.a.*) take a somewhat greater share than in the other genus, but they are still separated by a considerable interval from the subvertebral wedge-bone (*s.w.b.*) which constitutes the lower fourth of the cup. As in *Murænosaurus*, the lateral pieces of the neural arch do not unite above, but run back and articulate with the anterior zygapophyses of the axis; from the presence of a pair of peculiar facets on the exoccipitals in the young specimen (Pl. IX. fig. 4 *a, f.*), it seems probable that a pro-atlas was present, but whether single or paired there is no means of ascertaining. In the axis the neural arch (*ax.a.*) is lower than in *Murænosaurus*, and there is not so well-developed a neural spine; on either side of the base of the neural spine there is a strong ridge, not present in the other genus. The posterior zygapophyses (*p.z.*) are very large, and project a long way behind the level of the posterior face of the centrum; their facets are flat and look outwards and downwards. The posterior face of the centrum is concave in the middle with a raised and convex outer portion; it is considerably wider than high. The facet (*r.^{1}f.*) for the rib of the atlas is much larger than in *Murænosaurus*, and its antero-inferior portion is borne by the subvertebral wedge-bone. The rib of the axis (*r.^{2}*) is larger than that of the atlas; the facets for these two ribs are confluent. The subvertebral wedge-bone (*s.w.b.*) bears a strongly developed hypapophysial ridge (*hy.r.*), which forms a distinct anterior prominence not seen in *Murænosaurus*; the ridge only extends backwards on to the anterior portion of the centrum of the axis. In the young specimen (text-fig. 78, A, B) in which the constituent elements of the atlas-axis are still separable, there is no trace of the presence of a second subvertebral wedge-bone. In addition to the atlas and axis there are thirty other cervical vertebræ. Of these the centra are much shorter than in *Murænosaurus*, and their articular ends are strongly concave in the middle with rounded margins; in outline they are transversely oval, the upper border beneath the neural canal being a little concave. The ventral face of the centra is gently concave from before backwards and also from side to side, with the exception of the longitudinal ridge between the pair of nutritive foramina. Above the rib-facets the sides of the centra are flat or slightly concave. The neural arches occupy nearly the whole length of the centra. They bear very strongly developed anterior and posterior zygapophyses, the facets of which are nearly flat in the anterior part of the neck, but further back become somewhat concave and convex respectively from side to side. In the anterior cervical vertebræ the anterior and posterior zygapophyses are connected by a ridge which disappears on the posterior part of the neck. The

Text-fig. 78.

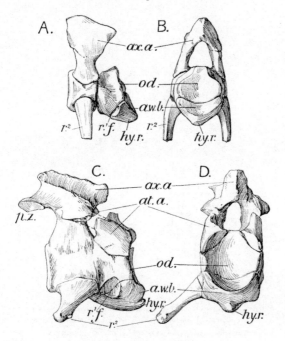

Atlas and axis of *Cryptocleidus oxoniensis*: A, from right side; B, from front of a young specimen (R. 2417, nat. size); C, from right side; and D, from front of an older specimen (R. 2860, nat. size).

at.a., arch of atlas; *a.w.b.*, anterior wedge-bone; *ax.a.*, arch of axis; *hy.r.*, hypapophysial ridge; $r.^2$, rib of axis; *r.¹f.*, facet for rib of atlas; *od.*, odontoid (centrum of atlas); *p.z.*, posterior zygapophysis of axis.

neural arches, like the centra, are much shorter from before back than in *Muraenosaurus*, and the neural spines seem never to have attained the height found in the posterior cervicals of some species of that genus. The facets for the cervical ribs are about as deep as long. In the anterior part of the neck they occupy nearly the whole length of the centrum, but further back are separated from the anterior border by a short interval. The cervical ribs are somewhat compressed from above downwards, and in some individuals in which ossification is very far advanced

Text-fig. 79.

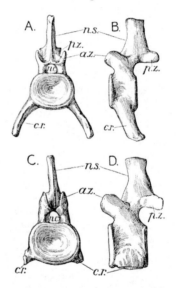

Anterior cervical vertebræ of *Cryptocleidus oxoniensis*: A and C, from front; B and D, from left side. (R. 2412, ½ nat. size.)

a.z., anterior zygapophysis; *c.r.*, cervical rib; *n.c.*, neural canal; *n.s.*, neural spine; *p.z.*, posterior zygapophysis.

(*e. g.*, R. 2862) they have a well-marked anterior angle; this is wanting in the nearly adult mounted specimen (R. 2860), in which the neural arches are fused with the centra throughout the column except in the posterior caudals, while the cervical and caudal ribs are free throughout. In the young skeleton (R. 2417) none of the arches and ribs are fused with the centra, and the cartilage-covered surfaces for the neura

Text-fig. 80.

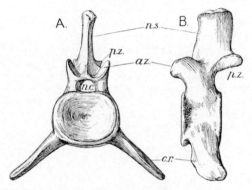

Posterior cervical vertebra of *Cryptocleidus oxoniensis*: A, from front;
B, from left side. (R. 2412, ½ nat. size.)

a.z., anterior zygapophysis; *c.r.*, cervical rib; *n.c.*, neural canal;
n.s., neural spine; *p.z.*, posterior zygapophysis.

Text-fig. 81.

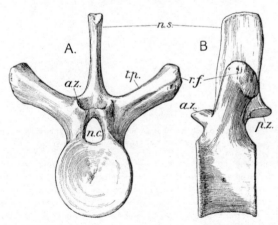

Middle dorsal vertebra of *Cryptocleidus oxoniensis*: A, from front; B, from left side.
(R. 2418, ½ nat. size.)

a.z., anterior zygapophysis; *n.c.*, neural canal; *n.s.*, neural spine; *p.z.*, posterior zygapophysis;
r.f., facet for rib; *t.p.*, transverse process.

pedicles are continuous with those for the cervical ribs. The neural spines are short and thickened; their summits were evidently cartilaginous. There are two pectoral vertebræ, in which the articulation for the rib passes from the centrum to the arch; their centra assume a more circular outline, passing into the form of the dorsal centra (text-figs. 81, 82), the vertical diameter of which is approximately equal to the transverse. At the same time the articular ends are less concave and the rounded rim disappears; the nutritive foramina pass on to the sides of the centra. There are 22 dorsals, the neural arches of which bear transverse processes; these increase in length and rise on the arch in the first seven or eight dorsals, while in the posterior five or six, on the other hand, they descend and shorten (text-fig. 82). In the middle of the back

Text-fig. 82.

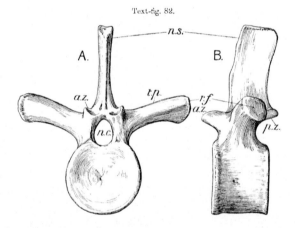

Posterior dorsal vertebra of *Cryptocleidus oxoniensis*: A, from front; B, from left side. (R. 2418, ½ nat. size.)

a.z., anterior zygapophysis; *n.c.*, neural canal; *n.s.*, neural spine; *p.z.*, posterior zygapophysis; *r.f.*, facet for rib; *t.p.*, transverse process.

they are moderately long and curved, the concavity being downwards. The neural spines on the dorsal region are shorter and narrower than in *Murænosaurus*; in the young animal they are stout, but much shorter than in the hinder part of the neck. The zygapophyses in the anterior part of the neck are larger, and look more directly upwards and downwards than in the posterior portion. There seem to have been four sacral vertebræ (text-fig. 83), each bearing a pair of stout ribs (*s.r.*), which articulate partly on the arch and partly on the centrum and are enlarged distally (*s.f.*); whether or not they actually joined the ilium is uncertain, but from the presence of deep roughened pits on the inner face of the upper end of the ilia in very old individuals, it seems possible that they did so, at least in advanced life. The centra in the sacrals

Text-fig. 83.

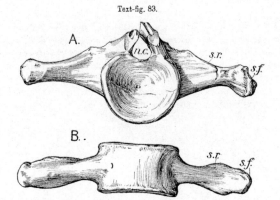

Sacral vertebra of *Cryptocleidus oxoniensis*: A, from front; B, from below. (R. 2412, ½ nat. size.)
n.c., neural canal; *s.f.*, facet for ilium; *s.r.*, sacral rib.

Text-fig. 84.

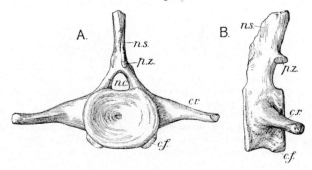

Anterior caudal vertebra of *Cryptocleidus oxoniensis*: A, from behind; B, from left side.
(R. 2412, ½ nat. size.)

c.f., facet for chevron-bone; *c.r.*, caudal rib; *n.c.*, neural canal; *n.s.*, neural spine; *p.z.*, posterior zygapophysis.

become slightly depressed, and pass posteriorly into the transversely oval centra of the anterior caudals (text-fig. 84). In the sacral region the neural spines begin to decrease in height, the decrease continuing to the end of the tail. The zygapophyses are well developed, somewhat concave (or convex) from side to side, and looking more inwards (or outwards) than further forwards. In the caudal region (text-figs. 84, 85) the centra, as already mentioned, are wider than deep; laterally, at least in the young, they bear prominent facets for union with the caudal ribs, which in the adult become joined to the centra. These ribs are compressed from above downwards, and a little behind the middle of the tail some are considerably expanded towards their outer ends; in all there is a tendency to curve backwards. The facets for the chevrons commence on the second or third caudal; at first they are confined to the hinder part of the centrum, where they form a pair of projections truncated by a flat nearly circular surface looking downwards and backwards. Further back (text-fig. 85) these facets are on both the

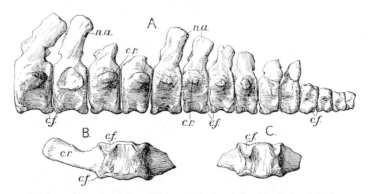

Posterior caudal vertebræ of *Cryptocleidus oxoniensis*: A, end of caudal series from left side; B and C, posterior caudal vertebræ from below. (R. 3705, ½ nat. size.)

c.f., facet for chevron-bone; *c.r.*, caudal rib; *n.a.*, neural arch.

anterior and posterior edges of the centra, the heads of the chevrons articulating between the successive centra; the posterior facet is the larger. The neural arches on the caudal region decrease in height gradually from before backwards; the zygapophyses disappear about the middle of the tail; the arches of the caudal vertebræ are the last to fuse with the centra, and the fusion takes place from before backwards.

The chevrons are not well known; in the young they are little rounded rods of bone with a slightly expanded vertebral extremity.

The cervical, sacral, and caudal ribs have already been referred to. The dorsal ribs are thickened and oval in section towards their articular ends, which terminate in a nearly flat oval facet for the transverse process. External to this proximal thickening they are somewhat compressed from before backwards, and their upper edge bears a sharp ridge which terminates externally in a slightly backwardly deflected crest. External to this again the ribs are nearly circular in section; they become very little thinner towards their lower end, which terminates in a flat or slightly concave surface probably tipped with cartilage in life.

The *ventral ribs* (text-fig. 86) form a close plastron consisting of eight or nine transverse rows of bones, each consisting of a median element (1) and three lateral pairs (2, 3, 4), besides two posterior rows in which the median element is wanting.

Text-fig. 86.

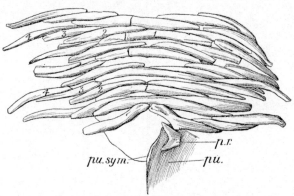

Ventral view of plastron of ventral ribs of *Cryptocleidus oxoniensis*. (R. 2862, ⅛ nat. size.)
pu., pubis; *p.r.*, forked end of posterior ventral rib; *pu.sym.*, symphysis of pubis;
1, 2, 3, 4, median and three lateral ribs of a transverse series.

The median bone of a row is oval or circular in section in its middle portion, which is often much thickened; towards the outer ends the bone thins down to a point, and its anterior face at either end bears a flattened facet for union with the inner end of the first lateral rib. This is pointed at both ends and forms a very open S-shaped curve; its inner end bears on its posterior face a surface for the overlap upon the median bone, while its outer end has on its anterior surface a facet for the reception of the inner end of the second lateral rib. This latter is similar to the first, and the third differs only in wanting the outer facet, the outer half being rounded or oval in section and terminating in a point. There is no trace of any connection with the true ribs. The two hinder rows differ from the others in not possessing a median element. In the anterior of the two the inner ends of the inner pair of lateral bones turn sharply forwards, and their extremities are closely adherent to the median bone of the row in front. The inner ends of the inner elements of the hindmost row are forked; what other lateral elements were present in this row is not known. This posterior row (*p.r.*) in the specimen figured seems to have actually underlain the anterior end of the pubis (*pu.*), but this may be the consequence of displacement

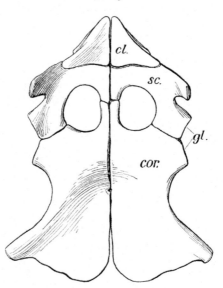

Adult shoulder-girdle of *Cryptocleidus oxoniensis*, from above. (R. 2616, about ⅛ nat. size.)
cl., clavicle; *cor.*, coracoid; *gl.*, glenoid cavity; *sc.*, scapula.

resulting from the flattening out of the carcass. The whole ventral surface of the body seems to have been very strongly protected, by the expanded scapulæ and coracoids in front, the plastron of ventral ribs in the middle, and the plate-like pubes and ischia behind.

Shoulder-girdle (Pl. X. figs. 1, 2; text-figs. 87–89).—This has been described in some detail in the Annals and Magazine of Natural History, [6] vol. xv. (1895) p. 333, and the following account is largely founded on the one there given.

CRYPTOCLEIDUS OXONIENSIS.

The *scapulæ* (*sc.*, Pl. X. figs. 1, 1 *a*, 1 *b*, 1 *c*; text-figs. 87–89) are, as usual in the group, triradiate bones, consisting of a backwardly directed bar carrying the articular surfaces for the coracoid and humerus, an upwardly directed process (*d.sc.*), and a ventral ramus (*v.sc.*), which in the adult extends forwards and inwards to the middle line, where it unites in symphysis with its fellow of the opposite side.

The following description of the scapula is based mainly on the adult shoulder-girdle (R. 2616) figured on Pl. X. and in text-fig. 87 :—

The posterior bar is triangular in section; its inner edge, forming the outer border of the coraco-scapular foramen, is sharp, thickening a little as it approaches the coracoidal surface; its upper outer border is rounded and passes above into the hinder border of the dorsal ramus or blade. The lower outer border rises into a rough ridge about 2 cm. from the glenoid surface and then runs forwards and outwards, forming on the outer face of the bone the boundary between ventral and lateral regions of the outer and ventral surfaces; anteriorly it terminates in a strong outwardly directed tubercle having a smooth facet on its summit. The glenoid surface and that for union with the coracoid are at right angles with one another, the line of junction being slightly concave and about 6·5 cm. in length. The form of the glenoid surface is that of half a rather irregular oval, measuring about 6·5 cm. from the middle of its line of union with the glenoid surface of the coracoid to the top of the curve. The surface for union with the coracoid is an isosceles triangle, the sides of which are slightly convex and measure 8·2 cm. in length: the base is the line of union with the glenoid surface. This latter is nearly smooth, while the coracoidal surface is greatly roughened by the presence of irregular pits and ridges.

The dorsal ramus of the scapula (*d.sc.*) is compressed from within outwards and is between 4 and 5 cm. wide at its summit, which is occupied by a rough depressed surface to which, in life, probably a small suprascapular cartilage was attached. The anterior border of this ramus, especially on its lower portion, is greatly roughened, probably for the attachment of muscles above, and at its lower end for union with the roughened facet on the outer angle of the clavicle.

The ventral ramus (*v.sc.*) is the largest and most important part of the scapula, at least in the adult; its anterior border is a continuation of the anterior edge of the dorsal ramus; it is at first rounded and concave as far as the prominent tubercle referred to above, then it becomes relatively sharp and thin, running inwards and forwards to the middle line, where by a sharp curve it passes into the median border with which it makes an angle of about 45°. The posterior border of this region of the scapula forms the anterior and half the inner edge of the coraco-scapular foramen; it increases in thickness from without inwards and then backwards to the point of union with the anterior prolongation of the coracoid (Pl. X. fig. 1 *c*), the surface for union with which is nearly semicircular and at right angles to the median symphysis. This latter (*sym.*), occupying the inner face of the thickened posterior prolongation of the ventral ramus, forms a nearly rectangular surface about 7–8 cm. long and 5 cm. deep; its antero-dorsal angle is rounded, while its antero-ventral angle is prolonged forwards as the inner edge of the thin anterior region of the scapula. The symphysial surface is deeply pitted and grooved by channels, which seem to have communicated with the exterior by a foramen situated on the upper surface at the middle of the symphysis. The outer surface of the ventral ramus of the scapula is nearly flat, but the visceral (upper) surface is divided into the raised and thickened symphysial region behind and the thin depressed anterior area which supports the clavicle, the two being separated by the ridge against which the hinder edge of the clavicle rests. In this thin anterior region the two scapulæ do not actually meet in the median line, but are separated by a narrow V-shaped interval, covered by the inner borders of the overlying clavicles; it is possible that this interval was filled by cartilage and that in advanced age the scapular symphysis was prolonged to the extreme anterior end.

The *clavicles* (*cl.*, Pl. X. figs. 1, 2; text-figs. 87–89) are in the form of scalene triangles. The outer and posterior borders meet in an acute angle and the posterior border is sometimes concave. The lower surface of the outer angle bears a roughened facet (*s.sc.*), which fits against a corresponding rugosity on the anterior edge of the scapula and no doubt united closely with it. This union of the outer ends of the clavicles with the anterior borders of the scapulæ is a point of considerable interest, since it is probably a remnant of the original condition seen in the Nothosauridæ (text-fig. 61 B, p. 108), in which the clavicular arch stretches from one scapula to the other, the ventral plates of these bones being still widely separated, as, indeed, they are in the young shoulder-girdle of *Cryptocleidus* (see below). The outer border is nearly straight and is thin and sharp, and is usually turned a little downwards so as to fit closely against the anterior edge of the scapula. The inner (median) borders (*sym.*) of the clavicles are thickened and meet in the middle line at least in the anterior third. Behind this there is a notch on each clavicle, which in a former paper was described as showing either that a blood-vessel passed between the two bones at this point or that a rudimentary interclavicle was present. The latter explanation seems to be the correct one, though it is by no means sure that this element always ossified. In the specimen figured on Pl. X. fig. 2 the notch is well developed, but the clavicles behind it seem to have met in the middle line, so that the interclavicle, if present at all, must have been very small. In the specimen shown in text-fig. 88, on the other hand, there is a distinct interclavicle, the anterior slightly forked end of which fits into the notches in the clavicles, while its thin posterior prolongation lies between them in the middle line for the posterior two-thirds of their length. The occurrence of this rudimentary interclavicle in *Cryptocleidus*, like the existence of rudimentary clavicles in *Murænosaurus*, shows that both these genera were probably derived from some form in which, as in *Tricleidus*, both clavicles and interclavicles were well developed.

The *coracoids* (*cor.*, Pl. X. fig. 1; text-figs. 87–89) are very large, and, so far as concerns their anterior region, massive bones. Their form will be best understood by reference to the figures. The anterior median prolongations of the coracoids towards the scapulæ are almost semicircular in section, the diameter of the semicircle being represented by the symphysial face. This portion of the bones forms a marked projection below the rest of the ventral face (Pl. X. figs. 1 *b*, 1 *c*). Behind the scapular processes the concave anterior border of each bone is thin and sharp, forming the hinder boundary of

Text-fig. 88.

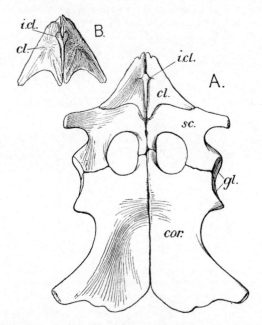

Adult shoulder-girdle of (?) *Cryptocleidus oxoniensis*, showing the rudimentary interclavicle:
A, from above; B, clavicles and interclavicle from below (outer surface). (R. 3538, about ½ nat. size.)

cl., clavicle; *cor.*, coracoid; *gl.*, glenoid cavity; *i.cl.*, interclavicle; *sc.*, scapula.

the coraco-scapular foramen. External to this the bone is greatly thickened and bears the facets for the scapula and the glenoid surface, these making an angle of about 135° with one another: the scapular facet is triangular and is greatly roughened by pits and ridges; the glenoid facet is half an oval, the short diameter being the line of junction with the scapular facet, its surface is gently concave and nearly smooth. Internal to these facets the bone is greatly thickened to the median symphysis, the form of which is shown on Pl. X. fig. 1 c. Behind this the bone thins greatly, especially towards the median line, and the straight symphysis is only interrupted at about a third of its length from the hinder end by a small foramen (*for.*) which is present in several specimens examined and probably transmitted a blood-vessel. The lateral borders, which are deeply concave, are somewhat thickened and terminate posteriorly in the posterolateral processes (*p.e.p*), which in adult individuals in this genus are strongly developed, and project outwards and backwards; they terminate in an oval concave surface, which was no doubt capped with cartilage during the life of the animal, and this cartilage

Text-fig. 89.

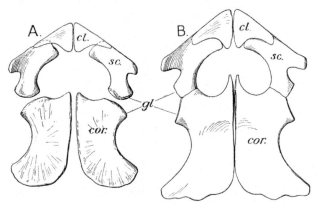

Immature shoulder-girdles of *Cryptocleidus oxoniensis*: A, a very young specimen with clavicles restored (R. 2416); B, an older example. (About ¼ nat. size.)

cl., clavicle; *cor.*, coracoid; *gl.*, glenoid cavity; *sc.*, scapula.

was continuous with the fringe of that substance which continued backwards the somewhat irregular posterior borders of the bones. The visceral surface of the united coracoids is concave from side to side, and, though slightly convex from before backwards in the region of the symphysial thickening, is strongly concave in that direction in its posterior two-thirds.

The above description refers to the shoulder-girdle of old individuals in which ossification is in an advanced condition. Fortunately the Leeds Collection contains shoulder-girdles of individuals of various ages, so that it has been possible to give an account* of the development of this part of the skeleton, and a short description of the growth-changes is appended.

In the young shoulder-girdle (text-fig. 89, A) the scapula is already triradiate in form, but the dorsal and, more particularly, the ventral rami are very imperfectly developed. The two bones did not meet in the middle line, and there is yet no trace of the extension backwards of the ventral rami to meet the anterior prolongations of the coracoids. The clavicles articulate by their outer ends with the anterior border of the scapulæ and meet in a median symphysis, the structure at this stage being essentially the same as in the primitive Sauropterygia. The ventral surface of the clavicles was exposed, the scapulæ, or at least their ossified portion, not yet extending beneath them. In the successively later stages the ventral rami of the scapulæ grow inwards and forwards beneath the clavicles and at the same time are gradually prolonged backwards in the middle line towards the gradually developing anterior median prolongations of the coracoids (text-fig. 89, B). Finally, the condition described above is attained, the scapulæ extending almost completely beneath the clavicles and meeting in a median symphysis, which through the backward continuation of the bone becomes continuous with the symphysis of the coracoids, the coraco-scapular foramina being completely separated from one another (text-fig. 87).

In the coracoids the chief growth-changes that take place are the prolongation forwards in the middle line to join the scapulæ, and the formation of the prominent postero-lateral processes (Pl. X. figs. 1 a, 1 b, p.e.p.; text-figs. 87, 88). As has already been pointed out, the prolongation inwards of the scapulæ beneath the clavicular arch causes the latter to become functionally unimportant or useless, and consequently in the family Elasmosauridæ it is extremely variable in form and is met with in all stages of reduction.

Fore Limb.—The fore paddle (text-figs. 90, 91 A) is chiefly remarkable for the great expansion of the distal end of the *humerus* in the adult: in the young this characteristic is not seen and only develops with advancing ossification (text-figs. 90, A–C). The head of the humerus is strongly convex in full-grown individuals, and its roughened surface shows that it was capped with cartilage. The tuberosity (*tu.*) is strongly developed and forms a quadrate prominence on the postero-superior surface at the upper end of the bone; its upper cartilage-covered surface is continuous with that of the head of the bone, or in individuals of advanced age separated from it by a slight concavity. The anterior border of the tuberosity is continued down as a strong ridge on the upper part of the shaft, while the posterior border forms a prominent angle continuous below with the posterior border of the shaft. The shaft is oval in section and increases gradually in width towards the distal expansion; its postaxial border

---

* "On the Development of the Shoulder-girdle of a Plesiosaur (*Cryptocleidus oxoniensis*, Phillips, sp.)," Ann. Mag. Nat. Hist. [6] vol. xv. 1895, p. 333.

bears a roughened surface for muscle-attachment, and its ventral surface towards its upper end is likewise roughened (text-fig. 90, C, *m.r.*). The great distal expansion bears two large facets for union with the radius and ulna: of these surfaces that for the radius is the larger and is slightly concave, while that for the ulna is usually straight or very gently concave; the two make a very obtuse angle with one another. In some cases there may have been a small postaxial accessory ossicle articulating with the humerus.

The *radius* (*r.*, text-figs. 90, 91) is very large and of peculiar form: this shape being already marked at a very early age (text-fig. 90, A, B) is of considerable value in determining as belonging to this genus very young individuals in which most of the

Text-fig. 90.

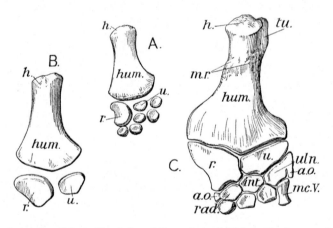

Proximal portions of fore paddles of *Cryptocleidus oxoniensis*: A, left fore paddle of very young individual (R. 2417, ⅓ nat. size); B, left fore paddle of older individual (R. 2416, ⅕ nat. size); C, right fore paddle (ventral face) in which ossification is complete (R. 2412, ¼ nat. size). In A the carpals are not in place.

a.o., accessory ossicle; *h.*, head of humerus; *hum.*, humerus; *int.*, intermedium; *mc.V.*, fifth metacarpal; *m.r.*, ridges for muscle-insertion; *r.*, radius; *rad.*, radiale; *u.*, ulna; *uln.*, ulnare; *tu.*, tuberosity of humerus.

other characteristics are not yet developed. In the adult the humeral border is slightly convex; its outer (preaxial) border is greatly elongated, convex above and concave below. The ulnar border is short and straight or slightly concave, while the distal border is also slightly concave and articulated with the radiale, with, in some cases, a short surface of contact with the intermedium. In one case (*a.o.*, text-fig. 91, A) it articulated externally by a short facet with a peculiar phalange-like bone lying on the preaxial side of the radiale (? a prepollex).

The *ulna* is much smaller than the radius; its humeral facet is gently convex, its inner (radial) surface is likewise convex, while distally it bears two flat facets making an obtuse angle with one another, for the intermedium and ulnare.

The proximal row of carpals consists essentially of the *radiale, intermedium,* and *ulnare,* but in many specimens there is a tendency to develop accessory ossicles on the preaxial or postaxial border or on both, and there is considerable variation in the form and manner of development of these accessory ossicles. In some cases it seems as if the increased width of the radiale and ulnare, consequent upon the expansion of the paddle, led to a tendency to ossify from more than one centre, and this may bring about the total or partial separation of the preaxial portion of the radiale and the postaxial part of the ulnare (text-fig. 90, C); often there is a want of symmetry in the paddles of opposite sides. In some cases it seems as if instead of a separation of the ulnare into two elements there has been a fusion with an originally distinct element (text-fig. 91, A, *a.o.*) corresponding to the bone called the pisiform in some of the paddles described above (e. g., *Tricleidus*, text-fig. 77, p. 160). The intermedium articulates mainly with the ulna, the facet for the radius being small and perhaps in some cases absent.

The distal carpals are three in number, the first (preaxial) articulating with the radiale (and anterior accessory ossicle if present), the second with the radiale and intermedium, the third (postaxial) with the intermedium and ulnare, while its postaxial border may have a facet at its proximal end for contact with the fifth metacarpal (text-fig. 91, A), which, as usual, articulates directly with the ulnare. The first metacarpal, which is flattened like the carpals, articulates only with the first distal carpal; the second metacarpal, which is cylindrical, with the second carpal; the third has two facets, one for the second the other for the third carpal, with which also the fourth metacarpal articulates; the fifth metacarpal has already been referred to. The *phalanges* are cylindrical and somewhat constricted in the middle; the faces by which they articulate with one another are nearly flat. The number of phalanges in the different digits is not known.

Pelvis (Pl. X. figs. 3, 3 *a*, 3 *b*; text-fig. 92).—The structure of the pelvis in this species has been described in some detail in Geol. Mag. [4] vol. iii. p. 145. It is composed of the usual three pairs of bones. The *pubis* (*pu.*, Pl. X. figs. 3, 3 *a*, 3 *b*) is relatively much wider from side to side and shorter from before backwards than is the case with the pubis of *Murænosaurus* (*cf.* text-fig. 65, p. 116). The anterior convex border is thin, except at its outer angle, where it is thickened and produced into a short process (*a.e.a*), separated by a notch from the median part of the border. This antero-external process of the pubis is also present in *Murænosaurus*, though less prominent; it has been suggested that it may be homologous with the lateral process of the Chelonian pubis.

The symphysial surface (Pl. X. fig. 3 b) is deeper and at the same time shorter than in *Murænosaurus*; its form will be best understood from the figure. The curvature of the symphysial border shows that even when ossification was far advanced, as in the specimen figured, the actual contact of the two bones, if present at all, must have been short, they being separated anteriorly and posteriorly with wedges of cartilage: the

Text-fig. 91.

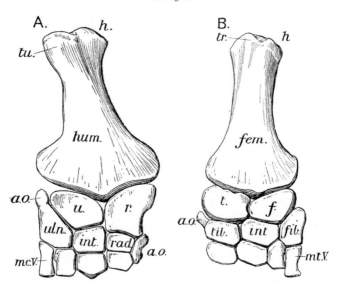

Proximal portions of fore and hind paddles of *Cryptocleidus oxoniensis*: A, right fore paddle (upper surface); B, left hind paddle (upper surface). (R. 2860, about ¼ nat. size.)

a.o., accessory ossicle; *f.*, fibula; *fem.*, femur; *fib.*, fibulare; *h.*, head of humerus and femur; *hum.*, humerus; *int.*, intermedium; *mc.V.*, fifth metacarpal; *mt.V.*, fifth metatarsal; *r.*, radius; *rad.*, radiale; *t.*, tibia; *tib.*, tibiale; *tr.*, trochanter of femur; *tu.*, tuberosity of humerus; *u.*, ulna; *uln.*, ulnare.

anterior cartilage was probably small and continuous with the cartilage bordering the front of the pubis; the posterior cartilage, on the other hand, was thick and probably was continuous with that uniting the symphysial surfaces of the ischia, thus completely separating the *foramina obturatoria* from one another.

The outer and posterior borders are both concave, thickening towards the massive articular region. This bears two facets, that for the ischium being nearly semicircular, and the diameter of the semicircle forming the line of division between it and the acetabular surface. This latter makes an angle of about 145° with the ischial surface and is slightly concave; there is no contact between the pubis and ilium.

The *ischium (isc.)* is of the usual hatchet-head form. The anterior portion of its symphysial border is thickened and bears a deep symphysial surface (Pl. **X.** fig. 3 *b*), behind which it thins rapidly, thickening again a little towards its posterior angle. The ischial symphysial surface was separated in front by a pad of cartilage, which, as already mentioned, was probably continuous with that between the pubes; probably there was also a small posterior cartilage. The neck of the ischium is comparatively narrow and depressed in section; towards the articular surfaces the bone becomes greatly thickened; there are three facets—one, looking almost directly forwards and semicircular in outline, for the pubis, a median one rectangular in outline and slightly concave forming the middle and greatest part of the acetabulum, and a posterior one looking backwards and outwards but only a little upwards, for the ilium, which slopes backwards much more than in *Murænosaurus*; the surfaces for the pubis and ilium are roughened for cartilage, the acetabular surface is smooth.

The *ilium (il.)* is a stout slightly curved rod of bone. Its lower end is greatly thickened and bears a nearly flat oblique oval surface, the inner two-thirds of which unite with the iliac facet of the ischium; the remaining third, making a slight angle with the rest, forms the posterior wall of the acetabulum. This surface was covered with cartilage, which also extended up on to an angular projection marked *c*. in the figure (Pl. **X.** fig. 3 *a*).

The middle part or shaft of the ilium is contracted and oval in section; it is curved, the concavity being anterior, and on the middle of its posterior border there is a small angular prominence probably for the attachment of muscles. The upper part of the bone is compressed and somewhat like the blade of an oar; the anterior border of this expanded region is thin and sharp, the posterior thick and rounded. The inner face of the upper end of the ilium is flat and, except in very old individuals, shows little or no trace of any union with the sacral ribs, to which probably it was attached loosely by ligaments.

In the pelvis, as a whole, it will be noted that the ilium is greatly inclined backwards, but this does not necessarily represent its exact inclination to the vertebral column, because probably the whole pelvis was inclined downwards and forwards. The pubes and ischia of opposite sides do not make a distinct angle with one another, their median portions being almost on the same plane, and the visceral surface of the ventral portion of the pelvis is only slightly concave from side to side. From before backwards the line of the symphysis is convex on the visceral side, its highest point being only about 8·5 cm. below a straight line joining the middle points of the acetabular cavities.

The pelvis of a young individual (R. 2417) is shown in text-fig. 92. None of the bones are completely ossified, but the immaturity of the pubis is most striking, there being no trace of the antero-external angle or of the articular surfaces; these must have been still cartilaginous. The ilium is less expanded at the ends, while the ischium, though resembling that of the adult more than is the case with the other elements, has its articular surfaces rounded and not sharply defined.

Hind Limb.—The *femur* (text-figs. 91 B, 93, *fem.*) is not greatly expanded at its distal end like the humerus. Its proximal end in fully ossified specimens bears a convex head, oval in outline, the surface of which is roughened and was covered with cartilage. The trochanter is large and prominent; its cartilage-covered upper end is continuous with that of the head; anteriorly and posteriorly it is marked off from the

Text-fig. 92.

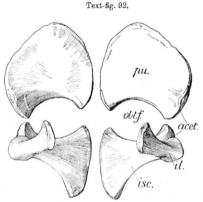

Immature pelvis of *Cryptocleidus oxoniensis*, from above. (R. 2417, ⅓ nat. size.)
acet., acetabulum; *il.*, ilium; *isc.*, ischium; *obt.f.*, obturator foramen; *pu.*, pubis.

upper part of the shaft by longitudinal grooves, of which the posterior is the more strongly marked; its outer surface is raised into ridges and roughened for muscle-attachment. The shaft is oval in section and bears on its ventral face and posterior border strong rugosities for muscle-attachment. The distal end, as already noted, is much less expanded than is the case with the humerus; it articulates only with the tibia and fibula, the facet for the first being slightly concave, that for the latter nearly flat.

The *tibia* (text-figs. 91 B, 93, *t.*) has a slightly convex femoral border; its preaxial edge is also convex; its inner (postaxial) border for union with the fibula is sometimes notched. Distally it bears a long facet for the tibiale and a shorter one, making an angle of about 130° with the last, for the anterior part of the intermedium. The *fibula* (text-figs. 91 B, 93, *f.*) has a long straight femoral edge, short and somewhat convex preaxial and postaxial borders, and distally two subequal facets for the inter-

Text-fig. 93.

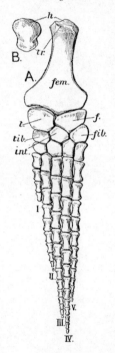

Left hind paddle of *Cryptocleidus oxoniensis*: A, upper surface; B, proximal end of femur.
(R. 3703, ⅓ nat. size.)

f., fibula; *fem.*, femur; *fib.*, fibulare; *h.*, head of femur; *int.*, intermedium; *t.*, tibia; *tib.*, tibiale; *tr.*, trochanter; I.–V., the five digits.

medium and fibulare, making an angle of about 120° with one another. The former of these facets is slightly concave. The *tibiale* (*tib.*) articulates proximally solely with

Text-fig. 94.

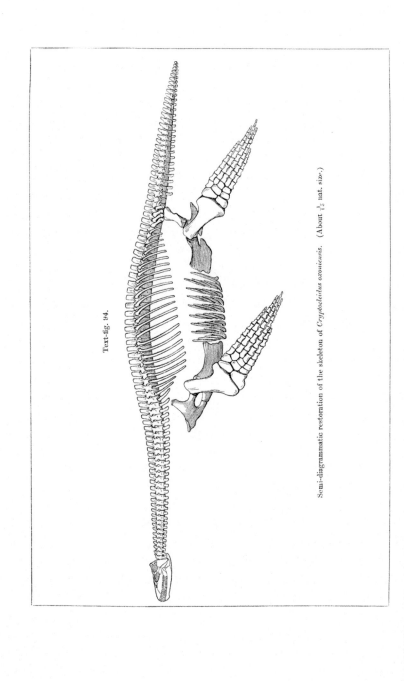

Semi-diagrammatic restoration of the skeleton of *Cryptocleidus oxoniensis*. (About $\frac{1}{15}$ nat. size.)

the tibia; distally it carries the first distal carpal and has a short surface for the preaxial part of the second. In some specimens (see text-fig. 91, B) there is a small accessory ossicle (*a.o.*) articulating with the preaxial border of the tibiale near its proximal end. The presence of this accessory ossicle may indicate that although the hind paddle is little expanded in comparison with the fore paddle, there is nevertheless a tendency towards such an enlargement. The *intermedium* (*int.*) has a short facet for the tibia and a longer one for the fibula, with which also the fibulare (*fib.*) articulates; the postaxial border of the last-mentioned bone is thin and convex. Of the three distal carpals, the first articulates with the tibiale alone, the second with the tibiale and intermedium, the third with the intermedium and fibulare, while its postaxial border is in contact with the fifth metatarsal, which, as usual, articulates directly with the fibulare. All the metatarsals are flattened, and the same is the case in a decreasing degree with the first two or three rows of phalanges. The more distal phalanges are more cylindrical and constricted in the middle till near the ends of the digits, where they become more flattened again; the terminal phalanges are mere nodules of bone. The articular surfaces of the phalanges are somewhat convex and the articulations between the successive phalanges of one digit usually alternate more or less regularly with those of the adjacent digits. In one nearly complete hind paddle (R. 3703, text-fig. 93) the numbers of the phalanges in the digits from the first to the fifth are 3, 9, 13, 13, 12.

A semi-diagrammatic restoration of the complete skeleton of *Cryptocleidus oxoniensis* is given in text-fig. 94. This drawing is made almost entirely from the skeleton of the adult (R. 2860) mounted in the Gallery of Fossil Reptiles, British Museum (see Frontispiece), and used as a basis for the description of the skeleton given above. The chief points of difference from *Murænosaurus* shown in this diagram are the relatively larger head, shorter neck, and more expanded fore paddles; the important differences in the shoulder-girdle and pelvis cannot be shown in a side view.

R. 2860 (Leeds Coll. 14). An almost complete skeleton of a nearly adult individual (Frontispiece). The skull is broken and incomplete, the parts preserved being:—basioccipital (Pl. IX. figs. 1, 1*a*), exoccipital, opisthotic (Pl. IX. figs. 1, 1 *a*), supraoccipital (Pl. IX. figs. 1, 1*a*), basisphenoid (Pl. IX. fig. 2), parietals, frontals, quadrate and squamosal (Pl. IX. fig. 3), premaxillæ (part), portions of pterygoids. The mandible is nearly complete. The vertebral column consists of the atlas and axis (text-fig. 78, C & D) and thirty other cervicals, two or three pectorals, twenty-one or twenty-two dorsals, three or four sacrals and twenty-two caudals, the distal portion of the tail being wanting; most of the vertebræ have their arches and ribs preserved: in the cervical and caudal regions the ribs in most cases have not yet fused with the centra. There are (as mounted) six rows of abdominal ribs, each consisting of a median element and three lateral pairs. The shoulder-girdle is complete; the fore paddles (text-fig. 91, A) want some of the phalanges; the pelvis is complete; the hind paddles (text-fig. 91, B) want the proximal end of the right femur, some tarsals, and phalanges. This specimen, which is mounted in the Gallery of

Fossil Reptiles, is probably the most nearly complete skeleton of an Elasmosaurian Plesiosaur known : the description of the skeleton given above and the restoration in text-figure 94 are founded mainly on it.

The dimensions (in centimetres) of this specimen are:—

Total length of the skeleton as mounted . . about (11 ft.)	335·0
Skull (Pl. IX. figs. 1–3):	
Length of basioccipital	3·5
Width of basioccipital at pterygoid processes	4·6
Transverse diameter of occipital condyle	2·4
Height from bottom of basioccipital to vertex of skull . .	8·6
Length of paroccipital processes . . . . . (approx.)	2·6
Width of articular surface of quadrate	3·1

Vertebræ	Atlas and axis (figured)	Fourth cervical	Tenth cervical	Fourteenth cervical	Twentieth cervical	Twenty-fifth cervical	Thirtieth cervical	First pectoral	Fifth dorsal	Tenth dorsal	Fifteenth dorsal	First sacral	First caudal	Twelfth caudal
Length of centrum in mid-ventral line	4·2	2·3	2·7	3·2	3·3	3·4	3·5	3·5	4·0	4·2	4·2	4·1	3·3	3·0
Width of posterior end of centrum.	2·8	2·8	3·4	3·9	4·5	5·0	...	...	...	...	...	...	6·1	4·7
Height of posterior end of centrum.	2·1	2·2	2·6	3·2	3·6	4·0	...	...	...	...	5·0	...	...	3·7
Height to top of neural spine .	5·1	6·1	7·6	8·9	10·3	11·0	...	...	15·5	...	14·8	11·9	12·2	8·9

Owing to the mounting of this skeleton, complete measurements of the vertebræ cannot be taken.

Shoulder-girdle :	
Greatest length of combined scapula and coracoid	52·1
,, ,, scapula	25·0
,, ,, coracoid . . . . . . (approx.)	39·0
Width of united coracoids at posterior angle of glenoid cavities	34·9
Width between outer ends of the postero-external angles of coracoids	48·0
Width of united coracoids at narrowest point	26·7
Length of clavicle (median border)	11·0
Width of clavicle (median border to outer angle)	10·6
Fore limb :	
Humerus : length	28·5
diameter of head	7·5
greatest width of upper end	10·9
width of shaft at narrowest	7·0
width of distal expansion	21·4
Radius : length of preaxial border	10·5
width of surface for humerus	10·9
Ulna : greatest length	5·6
,, width	9·3

Pelvis:
- Ilium: length ... 17·2
  - greatest width at upper end ... 7·3
  - ,,   ,,   lower end ... 5·6
- Pubis: greatest length ... 25·3
  - ,, width (from antero-external angle to symphysial border) ... 30·5
  - width of articular head ... 10·0
  - width between the antero-external angles of the two pubes ... 55·5
- Ischium: greatest width ... 18·1
  - width of articular head ... 8·8
  - length of expanded portion ... 18·4
  - width of neck ... 5·2

Hind limb:
- Femur: length ... 27·0
  - diameter of head ... 7·9
  - greatest width of upper end ... 9·5
  - width of shaft at narrowest ... 5·8
  - width of distal expansion ... 16·0
- Tibia: greatest length ... 5·7
  - ,, width ... 8·0
- Fibula: greatest length ... 5·1
  - ,, width ... 7·9

R. 2862 (Leeds Coll. 27). Imperfect skeleton of a large adult individual. The parts preserved are:—basioccipital, part of basisphenoid, atlas, axis and twenty-eight other cervical vertebræ, about twenty-three pectorals and dorsals and thirty sacrals and caudals; scapulæ, coracoids, clavicles, fore paddles wanting only a carpal and some distal phalanges, portions of ilia, ischia and pubes, hind paddles wanting some distal phalanges, ventral ribs (text-fig. 86) embedded in matrix and showing their relations to one another and to the pelvis. In a short interval between the ventral ribs and the pubes there is a peculiar wrinkled surface which may represent a portion of the abdominal wall.

This skeleton is that of a large and old individual, in which ossification is very far advanced. This is shown by the fact that all the neural arches, and, in the cervical and caudal regions, the ribs, are fused with the centra. The ossification of the limb-bones also is very far advanced, the heads of the humeri and femora being strongly convex, while the upper end of the tuberosity in the humerus and of the trochanter in the femur are much more clearly defined than in most specimens. In the shoulder-girdle all the sutures between the coracoids and scapulæ are obliterated, and the clavicles, which are very closely adherent to the visceral face of the scapulæ, are fused with one another in the middle line.

The specimen is also interesting as showing the exact arrangement of the elements of the ventral buckler (text-fig. 86). The general relations of the posterior rows of ventral ribs to the pelvis can also be made out, though some displacement may have taken place in the course of the flattening out that the carcass has undergone. In the present condition it appears that the posterior row of ventral ribs lay below the anterior portion of the pubes. It seems probable that in life they were just in front of them, and that the plastron with the shoulder-girdle and pelvis formed a complete bony armour on the ventral surface of the body. The form and arrangement of the ventral ribs with regard to one another in this specimen have been described above. In the fore paddles the ulnæ are very wide, and the postaxial portion of that on the left side, which is bent a little downwards, seems to be on the point of separating off from the main body as an accessory ossicle, traces of the line of division being clearly visible. On the right side this part of the bone is already separate, though it remains in very close contact with the remainder. Of course, these separate, or imperfectly separate, portions of the ulnæ may be interpreted also as accessory ossicles on the way to fusion with the main bone.

The dimensions (in centimetres) of this specimen are:—

Basioccipital: length			(approx.)	3·8
	width at pterygoid processes			5·4
	transverse diameter of condyle			2·6

Vertebræ	Atlas and axis.	Sixth cervical.
Length of centrum in mid-ventral line	5·0	3·2
Width of posterior end of centrum	3·4	4·0
Height ,, ,, ,,	2·5+	3·1+

The remainder of the cervicals and the other vertebræ are too much crushed to supply measurements of any value.

Humerus: length	34·1
long diameter of head	9·5
short ,, ,,	6·9
greatest width at upper end	12·8
width of shaft at narrowest point	7·6
width of distal expansion	24·8
Radius: length of preaxial border	13·3
,, humeral border	12·8
Ulna: length	7·2
width with postaxial portion	13·3
,, without postaxial portion	12·0

The pelvic bones are too imperfect to be measured.

Femur: length	31·3
diameter of head (exaggerated by crushing)	9·5
greatest width at upper end	10·2
width of shaft at narrowest	6·3
,, distal expansion	18·6
Tibia: greatest length	6·3
,, width	9·0
Fibula: greatest length	5·9
,, width	9·0
Length in mid-ventral line of plastron so far as preserved	24·5
Width (exaggerated by crushing)	78·0

R. 3730 (Leeds Coll. 144). Imperfect skeleton, including skull (imperfect posteriorly, Pl. IX. fig. 7), atlas, axis and fifteen other cervical vertebræ, the centra of the posterior caudal vertebræ still united with one another ; six ribs (cervical, dorsal, and sacral), left clavicle, radius, ulna, and the greater part of one fore paddle, together with a number of odd bones of the other ; distal halves of both femora, tibiæ, fibulæ, and the greater part of the other bones of both hind paddles.

The skull (Pl. IX. fig. 7) is very much broken and is incomplete posteriorly ; the upper and lower jaws are crushed together and it can be seen that the long sharp teeth of the upper and lower series alternated throughout : there are about 24–25 lower teeth on each side. In the cervical region the sutures between the centra and the neural arches and ribs are still open. The terminal caudals do not seem to have borne ribs, but the neural arches and chevrons were present on all but the last, or perhaps two last vertebræ, although in this specimen they are for the most part represented only by the facets for their attachment.

The skull is too much crushed to give any reliable measurements.

The dimensions (in centimetres) of other parts of this specimen are :—

Vertebræ.	Atlas and axis.	Anterior cervicals.		Middle cervical.
Length of centrum in mid-ventral line	4·2	2·6 app.	3·5	3·5 app.
Width of posterior face of centrum	3·0	3·1	4·5	5·3
Height of posterior face of centrum	..	2·3	5·5	4·1 app.

The length of the united centra of the ten posterior caudal vertebræ	14·9
Clavicle : length of outer border	21·3
,, symphysial border (a;p.)	13·1
Radius : length of preaxial border	14·1
,, humeral border	15·2
Femur : width of distal expansion	20·1
Tibia : length	6·5
width	9·2
Fibula : length	5·4
width	8·6

R. 2412 (Leeds Coll. 31). A great part of the skeleton of a large adult individual. The parts preserved are :—eighteen cervical vertebræ, mostly with fused arches and ribs (text-figs. 79, 80), two pectorals, a dorsal, two sacrals (text-fig. 83), and twenty-two caudals (text-fig. 84); ribs, abdominal ribs ; shoulder-girdle somewhat imperfect, especially posteriorly, and with only portions of the clavicles preserved ; right fore paddle (text-fig. 90, C), left ulna, incomplete pelvis wanting one ilium, hind paddles wanting some tarsals and phalanges. This specimen is the type upon which Professor Seeley founded the species *Cryptocleidus platymerus* : he figured the shoulder-girdle, clavicles, and fore paddle in Proc. Roy. Soc. vol. li. (1892) pp. 145–148, text-figs. 13–15, the genus (or, as Professor Seeley in some places calls it, subgenus) *Cryptocleidus* being founded for its reception. The shoulder-girdle differs from those of R. 2616 and R. 2860 in being a little narrower and more lightly built, although the size and degree of ossification of the humerus show that the animal was advanced in age; the differences, however, do not seem to be sufficiently great to warrant separation as a distinct species.

The dimensions (in centimetres) of this specimen are:—

Vertebræ	Anterior cervicals.			Posterior cervicals.			Pec-toral.	First dorsal.	Sacrals.		Anterior caudal.			Middle caudal.	Posterior caudal.
									fig.^{d}.		fig.^{d}.				
Length of centrum in mid-ventral line	2·4	2·8	3·1	3·5	3·8	3·8	..	4·8	4·1	4·7	3·9	3·6	3·5	3·1 3·0	2·8
Width of posterior face of centrum	3·3	3·8	4·2	4·6	5·7	6·3	7·5	7·1	7·0	7·0	6·9	7·4	7·4	5·6	4·7
Height of posterior face of centrum	2·5	2·9	3·3	3·5	4·2	4·7	5·7	5·8	5·5	5·5	4·9	5·2	5·4	4·5 3·7	3·1
Height to top of neural spine				..	..	..	9·1	11·2	12·7						

Distance between the outer ends of the sacral ribs 21·2 cm.

Shoulder-girdle (approximate only):
Scapula: greatest length . . . . 27·2
length of median border . . . . 19·0
Coracoid: length from middle of glenoid cavity to tip of
postero-external process . . . . 35·4
width of the united bones at the hinder angle of the
glenoid cavity . . . . 38·8
width of the combined bones at narrowest . . . . 30·3
Humerus: length . . . . 36·2
width of head . . . . 9·0
,, upper end with tuberosity . . . . . . 12·6
,, shaft at narrowest . . . . 7·8
,, distal expansion . . . . 27·0
Radius: length of preaxial border . . . . 14·5
,, humeral border . . . . 15·3
,, ulnar border . . . . (approx.) 4·5
Ulna: length . . . . 6·5
width . . . . 14·4
Pelvis:
Ilium: length . . . . 19·0
width of upper end . . . . . . (approx.) 4·9
,, lower end (crushed) . . . (approx.) 7·6
Pubis: width from anterior angle of acetabulum to
symphysis . . . . 26·7
width of articular head . . . . 12·5
Ischium: width from acetabular surface to symphysis . . . 24·3
width of neck . . . . 7·0
Femur: length . . . . 33·7
greatest width of head . . . . 9·7
width of upper end with trochanter . . (approx.) 10·8
,, shaft at narrowest . . . . 7·0
,, distal expansion . . . . 20·7

Tibia: length of preaxial border . . . . 7·8
   „   femoral border . . . . 10·2
Fibula: greatest length . . . . 5·7
   „   width . . . . 11·1

R. 2616 (Leeds Coll. 25*). Portions of a skeleton of a large adult individual. In this specimen the bones are quite uncrushed and not distorted: some are scored by deep scratches apparently made by the teeth of some predaceous reptile. The parts preserved are some cervical vertebræ, mostly with the neural arches, which are just becoming fused to the centra; one sacral and eight caudal vertebræ, the neural arches in most cases missing, the suture having remained open; some cervical ribs which had not yet united with the centra, and a number of abdominal ribs of great size and massiveness; nearly perfect shoulder-girdle and pelvis, and a few odd paddle-bones. The shoulder-girdle has been described and figured in Ann. Mag. Nat. Hist. [6] vol. xv. (1895) pp. 335–340, figs. 1, 2; also figured on Pl. X. figs. 1, 1 a, 1 b, 1 c. The pelvis described and figured in the Geol. Mag. [4] vol. iii. (1896) pp. 145–148; also figured on Pl. X. figs. 3, 3 a, 3 b.

The dimensions (in centimetres) of some parts of this skeleton are:—

Vertebræ	Posterior cervicals.			Sacral.	First caudal.	Middle caudal.
Length of centrum in mid-ventral line	3·7	3·8	3·9	3·7	3·8	3·5
Width of posterior end of centrum	5·3	5·5	6·7	7·2 app.	7·1	6·5
Height of posterior end of centrum	4·2	4·5	5·4	5·3 app.	5·3	5·0
Height to top of neural spine	11·5	11·8	16·0 app.	15·2	14·6	

Shoulder-girdle: greatest length	66·5
Clavicle: length of anterior border	18·0
„ symphysial border	12·1
Scapula: greatest length	32·2
length of median (symphysial) border	18·8
length in straight line from median border to top of dorsal ramus	24·5
length of glenoid surface . . . . . (approx.)	6·2
„ surface for coracoid	8·5
antero-posterior diameter of coraco-scapular opening	12·1
lateral diameter of coraco-scapular opening (approx.)	11·0
Coracoid: length	46·7
width of united coracoids at posterior angle of glenoid cavity	42·2

---

* The left scapula and clavicle were acquired by the Museum in 1892 and registered as R 1966, the remainder of the skeleton was received in 1895.

196 MARINE REPTILES OF THE OXFORD CLAY.

Coracoid: width of the united bones at narrowest	33·0
width between the outer ends of the postero-lateral processes	57·1
Pelvis: greatest length	46·8
width between the antero-external angles of the pubes	66·6
Ilium: length	20·0
width of upper end	6·2
,, lower end	8·0
Pubis: greatest length	25·8
,, width (from antero-external angle to symphysial border)	34·2
width of acetabular surface (approx.)	8·7
depth of acetabular surface	5·9
Ischium: greatest length of median expansion	21·9
width from acetabular surface to symphysis	22·3
width of articular head	10·7
,, neck	6·7
antero-posterior diameter of obturator foramen	10·2

R. 2417 (Leeds Coll. 36). A nearly complete skeleton of a very young individual. The parts preserved are:—skull (imperfect and much broken), the occipital portion and the basis cranii figured Pl. IX. figs. 4, 4 a, 5; mandible (Pl. IX. fig. 6) which carried 20–22 teeth on each side, atlas and axis (text-fig. 78, A, B), 30–31 other cervical vertebræ, 2–3 pectorals, 21–22 dorsals, 3–4 sacrals, and about 21 caudals; neural arches, cervical, dorsal, and caudal ribs, ventral ribs, chevrons; pectoral girdle and fore paddles (imperfect) (text-fig. 90, A), pelvis (text-fig. 92), and hind paddles (imperfect). This specimen was described and figured in the Geol. Mag. [4] vol. ii. p. 241, pl. ix. The ossification of the vertebral column is imperfect, the elements making up the atlas-axis complex being all free from one another, the neural arches nowhere fused with the centra, and all the cervical and caudal ribs free; the cartilage-covered surfaces for union with the arches and cervical ribs are continuous with one another. The ossification of the posterior caudal vertebræ seems even less advanced than it is further forwards in the column. In the shoulder-girdle the ventral ramus of the scapula is little developed, and probably did not extend at all beneath the clavicle; in the fore paddles there is little indication of the great distal expansion of the humerus characteristic of the adult, but the radius is already large and generally similar in form to that of the adult.

The dimensions (in centimetres) of this specimen are:—

Total length of the skeleton as mounted (about 6 feet)	184·0
Skull (Pl. IX. figs. 4, 4 a, 5):	
length from occipital condyle to tip of snout	19·3
,, of basioccipital	3·0
width of basioccipital at pterygoid processes	3·0
transverse diameter of occipital condyle	1·9
Mandible: length (approx.)	23·0
,, of symphysis (approx.)	2·2

CRYPTOCLEIDUS OXONIENSIS.

Atlas and axis (text-fig. 78, A, B):
  length . . . . 3·3
  width of posterior face of axis . . . . 2·1
  height to top of neural spine of axis . . . . 3·8

Vertebræ	Fifth cervical.	Tenth cervical.	Twentieth cervical.	Twenty-fifth cervical.	Thirtieth cervical.	Anterior dorsal.	Posterior dorsal.	Anterior caudal.	Middle caudal.
Length of centrum in mid-ventral line . .	1·6	1·7	2·0	2·1	2·1	2·4	2·4	1·9	1·6
Width of posterior face of centrum . .	2·4	2·3	3·6	3·9	4·2	4·4	4·1	4·2	3·4
Height of posterior face of centrum . .	1·7	2·0	2·5	3·0 (app.)	3·1	3·6	3·3	3·0	2·7
Height to top of neural spine . . .	4·2	4·8	6·1	7·6	7·4	7·6	6·6	6·2	4·2
Width between the ends of the transverse processes . .	. .	. .	. .	. .	. .	. .	8·3	6·8	

Shoulder-girdle:
  Clavicle: length of anterior border . . . . 10·3+
    „ symphysial border . . . . 6·7
  Scapula: greatest length . . . . 11·0
    width of articular end . . . . 4·9
  Coracoid: greatest length . . . . 17·3
    width of each at hinder angle of glenoid cavity . 11·0
  Humerus: length . . . . 17·5
    greatest width at upper end . . . . 5·9
    width of shaft (at narrowest) . . . . 4·9
    „ lower end . . . . 11·1
  Radius: length . . . . 5·5
    width . . . . 4·1
Pelvis (text-fig. 92):
  Ilium: length . . . . 10·3
    width of upper end . . . . 3·5
    „ middle of shaft . . . . 2·2
    „ distal end . . . . 3·5
  Pubis: length . . . . 14·2 (app.)
    width . . . . 13·0
  Ischium: width of articular head . . . . 4·5
    „ neck . . . . 2·9
    „ symphysial expansion . . . . 8·5
  Femur: length . . . . 18·0 (app.)
    width of upper end . . . . 4·6
    „ shaft (at narrowest) . . . . 4·3
    „ lower end . . . . 10·3

R. 2416 (Leeds Coll. 37). Portions of the skeleton of an immature individual in which ossification is very incomplete. The parts preserved are:—the centra of twenty-seven cervical vertebræ, two pectorals, fourteen dorsals, and three caudals; a few of the neural arches and cervical ribs, in all cases free from the centra; the scapulæ and coracoids (figured in Ann. Mag. Nat. Hist. [6] vol. xv. (1895) p. 341, fig. 3, B, with the clavicle restored, see also text-fig. 89, A), humeri, radii, ulnæ (text-fig. 90, B), pubes, ischia, and left femur.

The dimensions (in centimetres) of this specimen are :—

Vertebræ	Anterior cervicals.			Middle cervicals.		Posterior cervicals.	Pectoral.	Dorsals.		Caudal.
Length in mid-ventral line.	1·8	1·9	2·3	2·4	2·6	2·7	3·2	3·4	3·3	2·4
Width of posterior face of centrum.	2·5	2·8	3·3	3·7	4·3	4·7	4·8	4·9	4·9	4·5
Height of posterior face of centrum.	1·9	2·1	2·7	2·9	3·3	3·5	3·9	4·4	4·5	3·3

Shoulder-girdle (text-fig. 89, A):
Scapula: greatest length . . . . 15·9
width of articular head . . . . 6·7
,, neck . . . . 4·3
length in a straight line from the median angle to tip of dorsal ramus . . . . 16·5
Coracoid: length (from anterior internal angle to postero-external process) . . . . 26·4
width of each at level of posterior angle of the glenoid cavity . . . . 14·4
width (at narrowest) . . . . 11·0
Humerus (text-fig. 90 B): length . . . . 22·5
greatest width of upper end . . . . 7·9
width of shaft (at narrowest) . . . . 6·0
,, distal end . . . . 14·1
Radius (text-fig. 90, B): length of preaxial border . . . . 8·3
length of humeral border . . . . . . (approx.) 7·7
Ulna (text-fig. 90, B): length . . . . 5·0
width . . . . 6·5
Pelvis:
Pubis: greatest length . . . . 16·3
,, width . . . . 20·2
Ischium: width of articular head . . . . 6·5
,, neck . . . . 4·7
,, from acetabular to symphysial surface . . 13·2
,, of median expansion . . . . 12·5

R. 2431 (Leeds Coll. 38). Portions of the skeleton of a very young individual. The portions preserved include part of the parietals, centra of thirty-five dorsal and caudal vertebræ with a few detached arches, coracoids, right scapula, some bones of the fore paddles (including the characteristic radii), ilium, ischium, portions of pubes, femur. The ossification is very imperfect, the distal end of the humerus being no more expanded than that of the femur, from which it is distinguished with difficulty; the centra of the

CRYPTOCLEIDUS OXONIENSIS.

vertebræ are not yet united to the arches and caudal ribs, and in some cases are curiously compressed from above downwards, so as to be much wider in proportion to their height than usual: this may be in part the result of crushing, but probably also is characteristic in some degree of the very young centra of the posterior dorsals and caudals.

Some dimensions (in centimetres) of this specimen are:—

Coracoid: greatest length	13·0
width at posterior angle of glenoid cavity	9·3
Scapula: greatest length	9·7
length in straight line from median angle to tip of dorsal ramus	10·5
width of articular head	4·5
? Humerus: length	16·3
width of shaft	4·0
,, distal end	8·8
Ilium: length	9·1
width of upper end (approx.)	2·4
,, shaft	1·7
,, lower end	2·5
Ischium: width from acetabular surface to symphysis	9·7
,, of neck	3·0
,, of median expansion	8·3
,, of articular head	4·0
? Femur: length	16·0
width of shaft	4·3
,, distal end	8·9

R. 2420 (Leeds Coll. 43). Small set of bones belonging to an individual in which ossification was far advanced. The parts preserved are three anterior dorsal vertebræ, two middle caudal centra, left humerus, left femur, some odd paddle-bones, and portions of ribs.

The dimensions (in centimetres) of this specimen are:—

Vertebræ	Anterior dorsals.		
Length of centrum in mid-ventral line	3·8	4·2	4·1
Width of posterior face of centrum	6·4	6·3	6·3
Height of posterior face of centrum	5·3	5·5	5·6

Humerus: length	33·2
diameter of head	8·7
greatest width of upper end with tuberosity	13·0
width of shaft	7·9
,, distal expansion	24·5
Femur: length	30·2
diameter of head	8·9
greatest width of head with trochanter	10·5
width of shaft	6·7
,, distal expansion	17·8

R. 2418 (Leeds Coll. 40). Portion of the skeleton of a large adult individual. The parts preserved are two cervical vertebræ, two pectorals, eighteen dorsals (text-figs. 81, 82), eleven sacrals and anterior caudals, some portions of dorsal and caudal ribs, the left ilium. In the vertebræ the neural arches are already fused to the centra, and the same is the case with most of the cervical and caudal ribs. The neural spines are much abraded. The ilium is imperfect at its upper end.

The dimensions (in centimetres) of this specimen are:—

Vertebræ	Posterior cervical.	Pectoral.	Anterior dorsal.	Middle dorsals.		Posterior dorsal.	Sacral.	Anterior caudals.	
				figured.	figured.				
Length of centrum in mid-ventral line	4·0	4·8	4·8	5·3	4·9	4·7	4·2	4·2	3·7
Width of posterior face of centrum	6·5	5·7	6·2	6·4(app.)	6·3	6·9	7·0	6·7	6·0
Height of posterior face of centrum	5·2	5·8	5·3	6·2	5·7	5·5	5·2	5·3	4·7
Width between the outer ends of the transverse processes	..	..	14·0(app.)	15·3	..	10·7			

```
            Ilium: length . . . . . . . . . . . . . . . . . 24·0
                   width of middle of shaft . . . . . . . . .  4·2
```

R. 3538. A nearly complete and very slightly crushed shoulder-girdle of an adult individual. In this specimen there is a small interclavicle fitting in between the clavicles (text-fig. 88, A, B). The latter have the median border more elongated than usual and the posterior border is deeply concave. Anteriorly the clavicles project considerably in advance of the ventral rami of the scapulæ : the roughened surface at their outer angles for union with a corresponding surface on the anterior border of the scapulæ is well developed and fits closely against the opposing bone. The interclavicle is less regular in form than might be supposed from the figure ; the bifurcation at the upper end is asymmetrical, the left side being larger than the right ; the whole bone is curved somewhat with the concavity on the right side : these irregularities seem to result from the fact that the element is on the point of disappearing ; possibly it may have occurred with other shoulder-girdles of this species, having escaped notice owing to its small size and irregular form.

The dimensions (in centimetres) of this specimen are:—

```
    Shoulder-girdle: greatest length . . . . . . . . . . 65·5
        Clavicles: length of median (symphysial) border . . . . 14·9
               ,,  anterior border . . . . . . . . 19·0
               ,,  posterior border (in straight line) . . 11·2
        Interclavicle: length . . . . . . . . . . . . . 10·1
                       width at point of bifurcation . . . . . . 1·8
                       greatest width of anterior branch . . . . 1·5
        Scapula: greatest length . . . . . . . . . . . 28·8
                 length of median border . . . . . . . . 17·4
```

Scapula: length in a straight line from median border to tip
of dorsal ramus (exaggerated by crushing) . . 26·2
length of glenoid surface . . . . 5·4
,,   surface for coracoid . . . . 7·8
antero - posterior diameter of coraco - scapular
opening . . . . 11·5
lateral diameter of coraco-scapular opening . . . 10·4
Coracoid: length . . . . 42·7
width of united coracoids at posterior angle of
glenoid cavity . . . . 40·0
width of the united bones at narrowest . . . . 28·8

R. 2465. Left scapula of a half-grown individual. Noticed in Ann. Mag. Nat. Hist. [6] vol. xv. (1895) p. 340. In this specimen the backward prolongation of the median ramus had not yet reached the anterior prolongation of the coracoid, and in the middle line there must have been a wide **V**-shaped opening between the two scapulæ, so that the clavicles were still extensively exposed on the ventral surface.

R. 3705. A series of fourteen posterior caudal vertebræ of an adult individual. These specimens are shown in text-fig. 85. In the anterior vertebræ of the series the neural arch seems to have fused with the centrum, but although the caudal rib is firmly united, the suture does not seem to have been obliterated; further back both arch and ribs are free. Both the anterior and posterior pairs of facets for the chevrons are well developed, but the latter are the larger; they seem to continue to the extreme end of the tail, as also do the neural arches, while, on the other hand, probably the last four or five centra do not bear ribs.

The dimensions (in centimetres) of this series of vertebræ, numbering them from before backwards, are:—

Caudal vertebræ	1.	2.	3.	4.	5.	6.	7.	8.	9.	10.	11.	12.	13*, 14.
Length of centrum in midventral line	3·2	3·0	3·0	?	3·0	2·4	2·5	2·2	2·1	1·8	1·6	1·5	1·7
Width of posterior face of centrum	5·1	5·0	4·6	4·1	4·1	3·8	3·7	3·3	3·0	2·9			
Height of posterior face of centrum	4·2	4·1	3·9	?	3·6	3·5	3·0	2·7	2·5	2·4	..	..	1·4
Height to top of neural arch	8·5	..	..	..	7·4	6·9	5·8	5·0	..	4·1			

The last two or three centra are fused together, so that satisfactory measurements of them cannot be made. The terminal centrum is wanting.

R. 3703 (Leeds Coll. 174). Nearly complete left hind paddle (text-fig. 93).
The dimensions (in centimetres) of this specimen are:—
Total length of paddle . . . . 98·0
Femur: length . . . . 30·2
width of head . . . . 8·8

Femur: greatest width of upper end with trochanter . . . 10·0
       width of shaft (at narrowest) . . . . 6·2
       „   distal expansion . . . . 18·9
Tibia: length . . . . 7·0
       width . . . . 8·3
Fibula: length . . . . (approx.) 6·2
        width . . . . 9·2

INDEX.

[The asterisk denotes a figure on that page.]

Baptanodon, 2, 3, 4, 9, 16, 31, 33, 35, 48.
—— discus, 46.

Cimoliosaurus durobrivensis, 127.
—— eumerus, 164.
—— eurymerus, 164, 165.
—— plicatus, 120, 127.
—— snowii, 92.
Cryptocleidus, 164.
—— oxoniensis, 164.
    skull, 165.
    basioccipital, 166.
    basisphenoid, 166.
    exoccipital-opisthotic, 166.
    supraoccipital, 166.
    parietal, 167.
    frontal, 167.
    squamosal, 167.
    quadrate, 167.
    premaxilla, 167.
    mandible, 167.
    teeth, 167, 168.
    vertebral column, 168, 169 *, 170 *, 171 *, 172 *, 173 *, 174 *.
    ribs, 174.
    ventral buckler, 175 *.
    shoulder-girdle, 176 *, 179 *, 180 *.
    fore paddle, 181, 182 *, 184 *.
    pelvis, 183, 186 *.
    hind paddle, 184 *, 186, 187 *.
    restored skeleton, 188.
—— platymerus, 164, 165.

Dolichorhyncops, 113.

ELASMOSAURIDÆ, 77.

Hatteria, 13, 16, 82.

ICHTHYOPTERYGIA, 1.
Ichthyosaurus, 9, 14 *.
—— acutirostris, 24.
—— extremus, 54.
—— zetlandicus, 29.

Leptocheirus, 2.

Microdontosaurus, 2, 3.
Mixosaurus, 1.
—— cornalianus, 56.
Muraenosaurus, 4, 77.
    skull, 78, 85 *, 88 *.
    basioccipital, 78, 79 *, 80 *.
    basisphenoid, 79, 80 *.
    parasphenoid, 80 *, 81
    exoccipital, 80 *, 81, 83 *.
    opisthotic, 81, 83 *.
    prootic, 84.
    parietal, 84.
    frontal, 86.
    postfrontal, 86.
    postorbital, 87.
    squamosal, 86.
    jugal, 87.
    maxilla, 87.
    prefrontal, 87.

2 D 2

Murænosaurus (cont.).
    premaxilla, 88.
    vomer, 88.
    palatine, 89.
    pterygoid, 89.
    mandible, 89, 90 *.
    vertebral column, 92 *, 94 *, 95 *, 96 *, 98 *, 99 *, 100 *, 101 *, 102 *, 103 *, 104 *.
    ribs, 105, 106 *.
    shoulder-girdle, 106, 108 *, 109 *.
    fore paddle, 111, 112 *.
    pelvic girdle, 115, 116 *.
    hind paddle, 112 *, 117.
    restored skeleton, 118 *.
—— beloclis, 140, 144.
—— durobrivensis, 127.
—— leedsi, 120.
—— platyclis, 134.
—— plicatus, 120, 127.

Nothosaurus, 107.
    shoulder-girdle, 108 *.
    pelvic girdle, 114 *.

OPHTHALMOSAURIDÆ, 2.
Ophthalmosaurus, 2.
    skull, 4–35 *.
    basioccipital, 5 *.
    exoccipital, 6, 7 *.
    supraoccipital, 7 *.
    prootic, 9, 10 *.
    opisthotic, 9, 10 *.
    stapes, 10 *, 11.
    basisphenoid, 12, 13 *, 14 *, 15 *.
    parasphenoid, 15 *, 16.
    squamosal, 16, 17 *
    supratemporal, 18.
    quadrate, 18 *.
    quadrato-jugal, 19, 20 *.
    jugal, 20 *.
    postorbital, 17 *, 21.
    lachrymal, 21 *, 22.
    nasal, 22 *.
    maxilla, 23 *.
    premaxilla, 24.
    parietal, 24, 25 *, 26 *.

Ophthalmosaurus (cont.).
    frontal, 26 *, 27.
    postfrontal, 27 *.
    prefrontal, 28.
    pterygoid, 28, 29 *.
    palatine, 29 *.
    vomer, 30 *.
    sclerotic ring, 31.
    mandible, 31, 32 *, 33 *, 34 *, 35 *.
    dentition, 36.
    hyoid arch, 36.
    vertebral column, 36, 37 *, 39 *, 40 *, 41 *, 42 *, 43 *, 44 *, 45 *.
    ribs, 45 *.
    shoulder-girdle, 46, 47 *, 48 *, 50 *, 51 *.
    fore paddle, 49, 52 *, 55 *.
    pelvic girdle, 56, 57 *, 58 *, 59 *.
    hind paddle, 58, 60 *.
    restored skeleton, 62 *.
—— icenicus, 61.

Pantosaurus, 4.
Picrocleidus, 139.
—— beloclis, 140.
    skull, 141.
    basioccipital, 141.
    quadrate, 141.
    vertebræ, 141.
    shoulder-girdle, 140 *, 142.
    fore paddle, 143.
    femur, 144.
—— sp., 146.
PLESIOSAURIA, 77.
Plesiosaurus durobrivensis, 164.
—— eurymerus, 164.
—— leedsi, 120.
—— oxoniensis, 164.
—— plicatus, 120.
—— (?) rostratus, shoulder-girdle, 108 *.
Polycotylus, 91, 113.

Sauranodon, 2, 3.
SAUROPTERYGIA, 77.
Shastasaurus alexandræ, 47.
—— osmonti, 47.
Sphenodon, 156.

Thaumatosaurus arcuatus, 107.
Toretocnemus, 2.
Tricleidus, 91, 113, 149.
—— seeleyi, 149.
    skull, 149.
    basioccipital, 150, 151 *, 153 *.
    exoccipital, 150 *.
    opisthotic, 150 *.
    basisphenoid, 150, 151 *.
    parasphenoid, 151 *, 153 *.
    parietal, 152.
    frontal, 152.
    maxilla, 152.
    premaxilla, 152.

Tricleidus seeleyi (*cont.*).
    vomer, 152.
    pterygoid, 153 *, 154.
    palatine, 154.
    quadrate, 154, 155 *.
    squamosal, 155 *, 156.
    mandible, 156.
    teeth, 156.
    vertebral column, 156.
    shoulder-girdle, 157, 158 *.
    fore paddle, 160 *.
    pubis, 161.
    hind paddle, 161.
Trinacromerum, 91.

PLATE I.

Fig.		Page
1.	*Ophthalmosaurus icenicus*, Seeley; transverse section of root of tooth: magnified twelve diameters. [R. 3013.]	36
2.	Ditto; transverse section of crown of tooth: magnified twelve diameters. [R. 3013.]	36
3.	Ditto; transverse section across the base of the crown of tooth: magnified twelve diameters. [R. 3013.]	36
4, 5, 6.	Ditto; teeth: nat. size. [R. 3013.]	36
7, 8.	Ditto; anterior ends of the rami of the mandible of a young individual with teeth *in situ*: nat. size. [R. 2181.]	36
9.	Ditto; three sclerotic plates: one-half nat. size. [R. 2740.]	31
10.	Ditto; section across junction of two sclerotic plates showing the interlocking suture: nat. size. [R. 2740.]	31
11.	Ditto; left quadrate, outer face: two-thirds nat. size. [Type specimen, R. 2133.]	18
12.	Ditto; left stapes, from front: two-thirds nat. size. [Type specimen, R. 2133.]	11
13.	Ditto; basioccipital, from above: two-thirds nat. size. [Type specimen, R. 2133.]	5
14.	Ditto; basioccipital, from behind: two-thirds nat. size. [Type specimen, R. 2133.]	5
15.	Ditto; left opisthotic from above: two-thirds nat. size. [Type specimen, R. 2133.]	9

art., articular surface of quadrate.
boc.f., facet for basioccipital.
c., cement.
cond., occipital condyle.
d., dentine.
e., enamel.
exo.f., facet for exoccipital.
h.c., channel for horizontal semicircular canal.
n.c., floor of neural canal.
op.f., facet for opisthotic.

p.c., pulp-cavity of tooth.
p.e.a., postero-external angle of quadrate.
p.v.c., channel for posterior vertical semicircular canal.
q.f., facet for quadrate.
q.j.f., facet for quadrato-jugal.
r., ridge for muscle-attachment.
s., suture between sclerotic plates.
sq.f., facet for squamosal.
st.f., facet for stapes.

CATAL. MARINE REPT. OXFORD CLAY.   PLATE I.

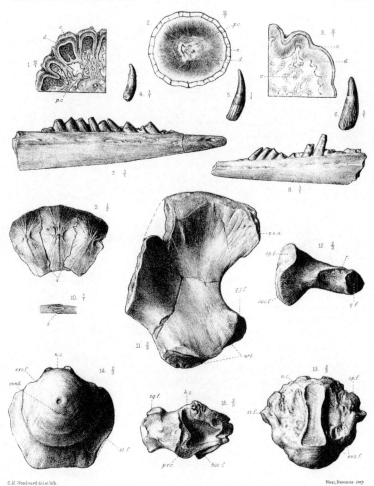

C.M. Woodward del.et lith.   West, Newman imp.

OPHTHALMOSAURUS.

PLATE II.

Fig.   Page

1. *Ophthalmosaurus icenicus*, Seeley; skull from above; the basioccipital is drawn from a second specimen: one-fourth nat. size.   [**R. 3702.**]   4

2, 2 *a*. Ditto; left prefrontal from above (2) and from below (2 *a*): one-third nat. size.   [Leeds Coll.]   28

3, 3 *a*. Ditto; postorbital, from outer side (3) and from inner side (3 *a*): one-third nat. size.   [Leeds Coll.]   21

4. Ditto; left fore paddle from above: one-fourth nat. size.   [**R. 3702.**]   49

These specimens were received after the descriptions were written, and therefore are not specially referred to in the text.

art., articular bone.	*p.*, pisiform.
b.oc., basioccipital.	*par.*, parietal.
fr., frontal.	*p.for.*, pineal foramen.
h., humerus.	*p.mx.*, premaxillæ.
int., intermedium.	*po.f.*, postfrontal.
j., jugal.	*po.f.s.*, surface for postfrontal.
j.s., surface for jugal.	*pr.f.*, prefrontal.
l., lachrymal.	*q.*, quadrate.
l.s., surface for lachrymal.	*r.*, radius.
mx., maxilla.	*rad.*, radiale.
n., nasal.	*st.*, stapes.
nar., external nares.	*u.*, ulna.
n.s., surface overlapped by nasal.	*uln.*, ulnare.

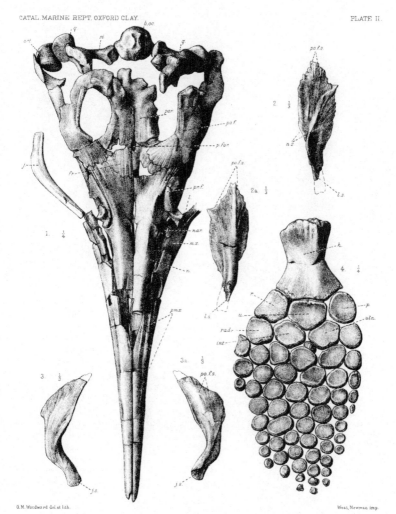

OPHTHALMOSAURUS.

PLATE III.

Fig.		Page
1, 1 a.	*Murænosaurus leedsi*, Seeley; basioccipital, exoccipital, and basisphenoid from below (1) and from side (1 a): half nat. size	78–82
2, 2 a.	Ditto; anterior part of skull from below (2) and from above (2 a): half nat. size	84–89
3, 3 a.	Ditto; mandible from above (3) and from below (3 a): half nat. size	89–92
4, 4 a.	Ditto; tooth, nat. size (4), and portion of crown, five times nat. size (4 a).	167
5, 6.	Ditto; teeth: nat. size	167

All the specimens figured in this Plate are parts of the type skeleton, **R. 2421**.

art., articular bone.
boc., basioccipital.
bs., basisphenoid.
dent., dentary.
exo., exoccipital.
fr., frontal.
mx., maxilla.

oc.c., occipital condyle.
pa., parietal.
pas., parasphenoid.
pin.f., pineal foramen.
pmx., premaxilla.
pt.f., facets for pterygoids.
spl., splenial.

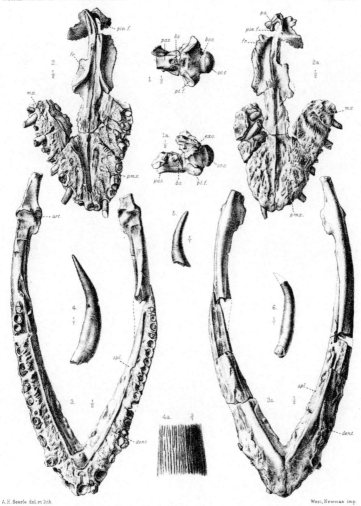

MURÆNOSAURUS LEEDSI.

PLATE IV.

Fig.		Page
1.	*Murænosaurus leedsi*, Seeley ; atlas and axis vertebræ from left side : one-third nat. size	92
2, 2 a.	Ditto; anterior cervical vertebra from right side (2) and from behind (2 a): one-third nat. size	94
3, 3 a.	Ditto ; middle cervical vertebra from right side (3) and from behind (3 a): one-third nat. size	95
4, 4 a.	Ditto; posterior cervical vertebra from right side (4) and from behind (4 a): one-third nat. size	95
5, 5 a.	Ditto ; anterior dorsal vertebra from right side (5) and from front (5 a): one-third nat. size	97
6, 6 a.	Ditto; anterior caudal vertebra from behind (6) and from below (6 a): one-third nat. size	102
7.	Ditto ; left fore paddle from above : one-fifth nat. size	111
8, 8 a.	Ditto ; left ilium from outer (8) and inner (8 a) sides : one-third nat. size.	115
9.	Ditto ; left ischium, upper (visceral) surface : one-third nat. size	117
10.	Ditto ; right hind paddle from above : one-fifth nat. size	117

All the specimens figured in this Plate are parts of the type specimen, **R. 2421.**

acet., acetabular surface.
at., atlas vertebra.
ax., axis vertebra.
a.z., anterior zygapophyses.
ch., chevron.
ch.f., facet for chevron.
c.r., caudal rib.
cr.i., crista ilii.
f., fibula.
fem., femur.
fib., fibulare.
h., humerus.
il.f., facet for ilium.
int., intermedium.

is.f., facet for ischium.
n.sp., neural spine.
pu.f., facet for pubis.
p.z., posterior zygapophyses.
r., ribs (in fig. 7, radius).
rad., radiale.
r.f., facet for rib.
s.f., sacral facet.
sym., symphysial surface of ischium.
t., tibia.
tib., tibiale.
t.p., transverse process.
u., ulna.
uln., ulnare.

CATAL. MARINE REPT. OXFORD CLAY. PLATE IV.

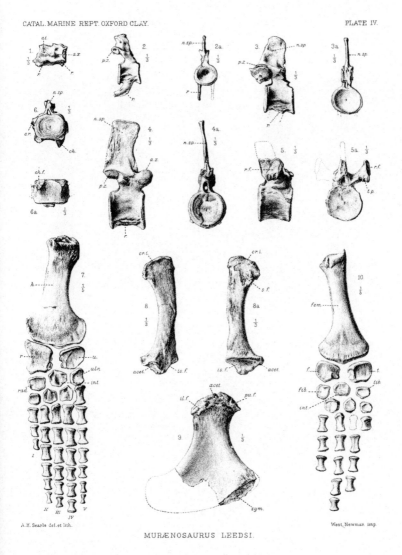

MURÆNOSAURUS LEEDSI.

PLATE V.

Fig.		Page
1, 1 a, 1 b.	*Murænosaurus durobrivensis*, Lydekker; centrum of anterior cervical vertebra from behind (1), from right side (1 a), from below (1 b): one-third nat. size	129
2, 2 a, 2 b.	Ditto; centrum of middle cervical vertebra from behind (2), from left side (2 a), from below (2 b): one-third nat. size	129
3, 3 a, 3 b.	Ditto; centrum of posterior cervical vertebra from behind (3), from left side (3 a), from below (3 b): one-third nat. size	129
4.	Ditto; pectoral vertebra from right side: one-third nat. size	129
5.	Ditto; centrum of dorsal vertebra from end: one-third nat. size	129
6, 6 a, 6 b.	Ditto; centrum of anterior caudal vertebra from behind (6), from left side (6 a), from below (6 b): one-third nat. size	129
7, 7 a.	Ditto; centrum of posterior caudal vertebra from behind (7), from below (7 a): one-third nat. size	129
8, 8 a, 8 b.	Ditto; right ilium, inner face (8), outer face (8 a), acetabular end (8 b): one-third nat. size	130
9, 9 a.	Ditto; right ischium, inner (visceral) face (9), acetabular end (9 a): one-third nat. size	130
10.	Ditto; ventral face of interclavicle: one-third nat. size	131
11, 11 a, 11 b.	Ditto; right humerus, ventral side (11), dorsal side (11 a), proximal end (11 b): one-fifth nat. size	130
12, 12 a, 12 b.	Ditto; right femur, tibia, and fibula, ventral side (12), dorsal side (12 a), proximal end of femur (12 b): one-fifth nat. size	130

All the specimens figured in this Plate, with the exception of the interclavicle (fig. 10), are parts of the type skeleton, R. 2428. The interclavicle (fig. 10) is from R. 2863.

acet., acetabular surface of ilium and ischium.
a.f., facet for pedicle of neural arch.
a.n., anterior notch of interclavicle.
c., cartilage-covered crest of acetabular end of ilium.
c.f., facet for chevron-bone.
cr.i., crista ilii.
f., fibula.
h., head of humerus and femur.
il.f., facet for ilium.
is.f., facet for ischium.
l.p., tuberosity of humerus.

m.r., ridges for muscle-attachment on the ventral faces of the humerus and femur.
p.n., posterior notch of interclavicle.
pu.f., facet for pubis.
ra.f., facet for radius.
r.f., facet for rib.
s.f., surface to which outer ends of sacral ribs were probably united.
sym., symphysial surface of ischium.
t., tibia.
t.p., transverse process.
tr., trochanter of femur.
u.f., facet for ulna.

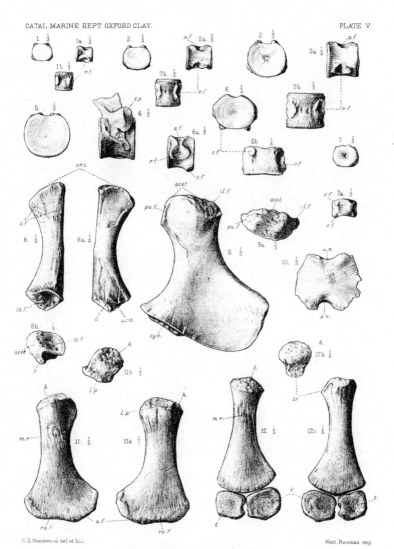

MURÆNOSAURUS DUROBRIVENSIS.

PLATE VI.

Fig.   Page

1. *Murænosaurus platyclis*, Seeley; skull from above: one-third nat. size.   [Type specimen, R. 2678.]   136

2. Ditto; mandible from above: one-third nat. size.   [Type specimen, R. 2678.]   136

3. Ditto; anterior portion of shoulder-girdle from above: one-fourth nat. size.   [Type specimen, R. 2678.]   136

4. Ditto; posterior cervical vertebra from right side: one-third nat. size.   [Type specimen, R. 2678]   135-6

5, 5 a, 5 b. Ditto; anterior cervical vertebra from below (5), from left side (5 a), and from front (5 b): one-third nat. size.   [Type specimen, R. 2678.]   135-6

6, 6 a. *Murænosaurus leedsi*, Seeley; interclavicle from below (6) and from above (6 a): one-fourth nat. size.   [R. 3704.]   126

a.n., anterior notch of interclavicle.
a.z., anterior zygapophysis.
boc., basioccipital.
cl., clavicle.
cor., coracoid.
d., dentary bone.
d.sc., dorsal ramus of scapula.
f., frontal.
gl., glenoid cavity.
i.cl., interclavicle.
i.s.f., interscapular fenestra.
j., jugal.
mx., maxilla.
nar., external nares.
n.sp., neural spine.

par., parietal.
par.p., paroccipital process.
p.f., pineal foramen.
pmx., premaxilla.
po.f., postfrontal.
p.orb., postorbital.
p.p., posterior process of interclavicle.
pr.f., prefrontal.
p.z., posterior zygapophyses.
r., cervical rib.
s.ang. & art., united surangular and articular.
spl., splenial.
sq., squamosal.
sym., mandibular symphysis.
v.sc., ventral ramus of scapula.

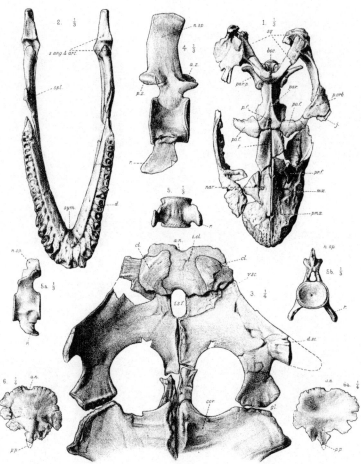

1–5, MURÆNOSAURUS PLATYCLIS.
6,   ,,   LEEDSI.

PLATE VII.

Fig.   Page

1. *Picrocleidus beloclis*, Seeley, sp.; posterior portion of right ramus of mandible: one-half nat. size.   [Type specimen, **R. 1965**.]   141

2, 2 a, 2 b. Ditto; shoulder-girdle (2) and proximal portion of left fore paddle (2 b) from above: one-fourth nat. size; interclavicle (2 a) from below: one-half nat. size.   [Type specimen, **R. 1965**.]   142

3. Ditto; posterior cervical vertebræ from right side: one-half nat. size.   [Type specimen, **R. 1965**.]   141

4. Ditto; posterior cervical vertebra from front: one-half nat. size.   [Type specimen, **R. 1965**.]   141

5, 5 a, 5 b. Ditto; cervical vertebræ from right side (5); front views of the vertebræ marked a and b (5 a, 5 b): one-half nat. size.   [**R. 3698**.]   141

6, 6 a. Ditto; caudal vertebræ from right side (6) and from front (6 a): one-half nat. size.   [**R. 3698**.]   141

a, *b*, vertebræ in figure 5 shown from front in figs. 5 *a*, 5 *b*.
ang., angular bone.
a.p., anterior process of cervical rib.
a.z., anterior zygapophysis.
c.f., facets for chevrons.
cor., coracoids.
d., dentary bone.
d.sc., dorsal ramus of scapula.
gl., glenoid cavity.
h., head of humerus.

i.cl., interclavicle.
␣l.p., tuberosity of humerus.
n.sp., neural spine.
p.z., posterior zygapophyses.
r., cervical ribs; in fig. 2 *b*, radius.
r^1., r^2., ribs of atlas and axis respectively.
r.f., facets for ribs.
s.ang. & art., fused surangular and articular bones.
u., ulna.
v.sc., ventral ramus of scapula.

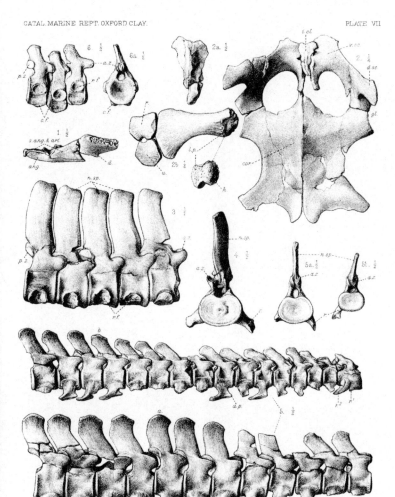

PICROCLEIDUS BELOCLIS.

PLATE VIII.

Fig.		Page
1, 1 a.	*Tricleidus seeleyi*, Andrews; right half of the mandible, from outer side (1) and from inner side (1 a): one-half nat. size	156
2.	Ditto; lower tooth : nat. size	156
3.	Ditto; clavicular arch, from above (the interclavicle has been broken longitudinally, but the crack is not drawn) : one-half nat. size	158
4, 4 a.	Ditto; left femur and proximal part of paddle, from below (4) and showing upper end of femur (4 a): one-third nat. size	161
5, 5 a.	Ditto; dorsal vertebra, from front (5) and from right side (5 a) : one-third nat. size	157
6, 6 a.	Ditto; dorsal vertebra, from front (6) and from right side (6 a) : one-third nat. size	157
7, 7 a.	Ditto ; pectoral vertebra, from front (7) and from right side (7 a) : one-third nat. size	157
8, 8 a, 8 b, 8 c, 8 d.	Ditto; series of cervical vertebræ from right side (8), the vertebræ marked a, b, c, d being shown from the front in figs. 8 a, 8 b, 8 c, 8 d : one-third nat. size. . .	156–7

All the above figures are from the type specimen, R. 3539.

a, b, c, d, in fig. 8 mark the vertebræ figured from the front in figs. 8 a, 8 b, 8 c, 8 d.
ang., angular bone.
at.a., neural arch of atlas.
ax.a., neural arch of axis.
cl., clavicles.
d., dentary bone.
f., fibula.
h., head of femur.
i.cl., interclavicle.
int., intermedium.
m.r., ridges for the attachment of muscles.

n.sp., neural spine.
r., cervical ribs.
r^1., r^2., ribs of atlas and axis respectively.
r.f., facet for rib.
s.ang. & art., fused surangular and articular bones.
spl., splenial.
sym., symphysis of mandible.
t., tibia.
tib., tibiale.
t.p., transverse process.
tr., trochanter of femur.

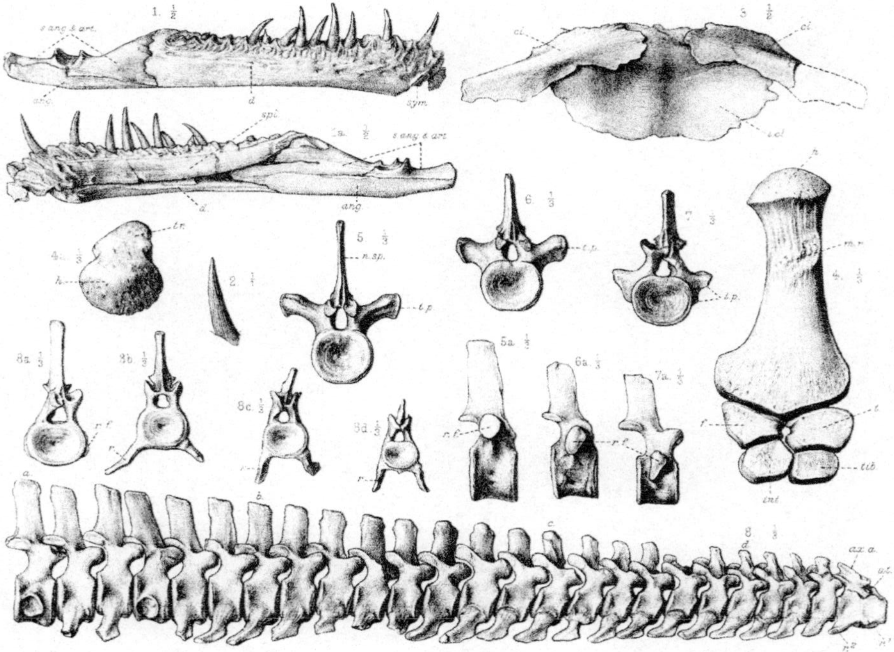

CATAL. MARINE REPT. OXFORD CLAY.

PLATE VIII.

TRICLEIDUS SEELEYI.

PLATE IX.

Fig.		Page
1, 1 a.	*Cryptocleidus oxoniensis*, Phillips, sp.; occipital region of skull from front (1), from behind (1 a): two-thirds nat. size. [R. 2860.]	166
2.	Ditto; basis cranii from below: two-thirds nat. size. [R. 2860.]	166
3.	Ditto; right squamosal and quadrate: two-thirds nat. size. [R. 2860.]	167
4, 4 a.	Ditto; occipital region of skull of a young individual, from front (4) and from behind (4 a): two-thirds nat. size. [R. 2417.]	166
5.	Ditto; basis cranii of a young individual from below: two-thirds nat. size. [R. 2417.]	166
6.	Ditto; mandible from above: two-thirds nat. size. [R. 2417.]	167
7.	Ditto; crushed anterior portion of skull and mandible from below: two-thirds nat. size. [R. 3730.]	166–7

a., ampulla of posterior vertical semicircular canal.
art., articular surface for quadrate.
boc., basioccipital.
bsp., basisphenoid.
bsp.f., facet for union with the basisphenoid.
d., dentary bone.
ex.op., united exoccipital and opisthotic bones.
f., facet on hinder face of exoccipital, possibly for pro-atlas.
for., median opening in basisphenoid of young individual.
h.c., channel for horizontal semicircular canal.
i.c.f., internal carotid foramen.
jug., jugular foramen.
l.p., lateral process of parietal.
pal., palatine.
par., parietal.
par.p., paroccipital process.
pas., parasphenoid.
pt.p., pterygoid (lateral) processes of basioccipital.
p.v.c., channel for posterior vertical semicircular canal.
q., quadrate.
soc., supraoccipital.
spl., splenial.
sq, zygomatic process of squamosal.
sq.', parietal process of squamosal.
sym., symphysis of mandible.
XII., foramen for twelfth nerve.

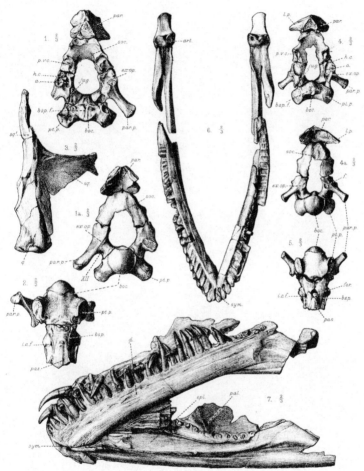

CRYPTOCLEIDUS OXONIENSIS.

PLATE X.

Fig.   Page

1, 1 *a*, 1 *b*, 1 *c*. *Cryptocleidus oxoniensis*, Phillips, sp.; shoulder-girdle from above (1), from front (1 *a*), from left side (1 *b*), and showing symphysial surfaces (1 *c*): one-sixth nat. size.   [**R. 2616.**]   170

2. 2 *a*. Ditto; left clavicle from upper (visceral) side (2) and from ventral (scapular) side (2 *a*): one-sixth nat. size.   [**R. 2616.**]   178

3, 3 *a*, 3 *b*. Ditto; pelvis from above (3), from right side (3 *a*), and showing symphysial surfaces (3 *b*): one-sixth nat. size.   [**R. 2616.**]   183

acet., acetabulum.
a.e.a., antero-external angle of pubis.
  c., ridge on ilium above acetabular surface.
  cl., clavicle.
  cor., coracoid.
  cr.i., the crest of the ilium.
  d.sc., dorsal ramus of scapulæ.
  for., foramen between coracoids.
  gl., glenoid cavity.

il., ilium.
isch., ischium.
obt.f., obturator foramen.
p.e.p., postero-external process of coracoid.
pu., pubis.
sc., scapula.
s.sc., surface on clavicle for union with scapula.
sym., median symphysis between various bones.
v.sc., ventral ramus of scapula.

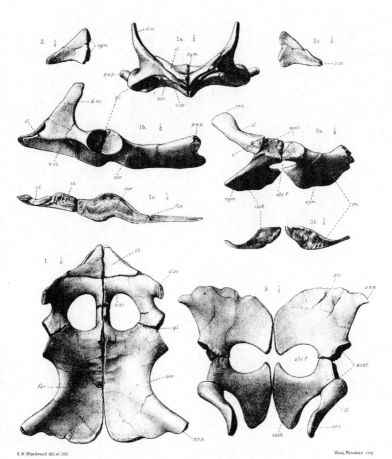

CRYPTOCLEIDUS OXONIENSIS.

Printed in Great Britain
by Amazon